Madison A. Cooper

Eggs in Cold Storage

Theory and Practice in Preserving Eggs by Refrigeration

Madison A. Cooper

Eggs in Cold Storage
Theory and Practice in Preserving Eggs by Refrigeration

ISBN/EAN: 9783744713917

Printed in Europe, USA, Canada, Australia, Japan

Cover: Foto ©berggeist007 / pixelio.de

More available books at **www.hansebooks.com**

THEORY AND PRACTICE IN PRESERVING EGGS BY RE-
FRIGERATION. DATA, EXPERIMENTS, HINTS ON
CONSTRUCTION, ETC., FROM PRACTICAL
EXPERIENCE, WITH ILLUSTRATIONS.

BY

MADISON COOPER.

CHICAGO:
H. S. RICH & CO.
1899.

PREFACE.

IN the interest of a better understanding and dissemination of knowledge on the cold storage of eggs, the writer has communicated with quite a large number of individuals and companies, asking their ideas and requesting that they give full answers to a printed list of questions sent them. Although, at first, the replies were rather slow in coming in, the total result of these letters has been most gratifying; nearly one-half acknowledging receipt of the inquiry, and more than one-half of this number giving fairly full replies to the questions submitted. Considering the fact that the inquiries were regarded by some as being of a rather personal nature, the proportion of managers sending full replies is large. Several gentlemen were frank enough to say that personal considerations prevented them from giving any information; others gave guarded or partial replies. In the main, however, storage men have been willing to give information and exchange ideas.

The list of inquiries sent out covers the subject very thoroughly, and divides it into six different parts, with three separate questions relating to each. To the data so cheerfully furnished by others is added information from the writer's experience and practice, with such explanation of theory and practice as may seem necessary to a clear understanding of the principles of successful egg refrigeration. It is hoped that those who are new to the business may obtain valuable information from these collected data, and that those with experience may derive some benefit in the way of a review, and possibly pick up some new ideas as well.

A large portion of the matter contained in these

pages appeared in *Ice and Refrigeration* as a series of articles entitled: "Eggs in Cold Storage." The present book is printed for the purpose of putting the matter in permanent form, believing that those who have followed the original articles would find it convenient for future reference. While the present book has many shortcomings, and there is no doubt room for the addition of much information, reliable data, and the results of extended observations and tests, there has not heretofore been anything like a complete write-up of the subject; and in consideration of this fact the reader is asked to be liberal in his criticism.

If any errors or lack of details are noted, the author would gladly acknowledge and explain the points at fault if his attention is called to any. No other object has been in mind in writing these articles than a furtherance of scientific knowledge on the subject of refrigeration as applied to the preservation of perishable products, and the great assistance rendered by those who have written painstaking replies to the list of inquiries is hereby acknowledged. The combination and comparison of information are beneficial, and if those who have further data or records of tests will only put them before others in their line of business, no loss will be sustained by the individual giving the information, while much general good will result.

INTRODUCTION.

THE value of the eggs placed in cold storage for preservation is estimated at about $20,000,0U0 annually for the United States alone. Considering the importance the industry has already attained, its rapid growth and future outlook, the amount of accurate information available to those engaged in the business seems very meager. The difficulties to be overcome, the skill required,and the importance of a well designed structure are not usually explained by those interested in promoting new enterprises in this line, and consequently not appreciated by those making the investment. Financial disaster has overtaken many large companies who have erected costly refrigerating warehouses; those which have succeeded have been forced to install new systems, make expensive changes, and make a thorough study of the products handled. The experience of nearly all has been emphasized at times by heavy losses paid in claims made by customers for damage to goods while in storage, or the necessity of running a large house while doing a very small business. Those about to become interested in the business may find food for thought in the above, and the history of a dozen houses, in different localities,will be good information for would-be investors.

The scarcity of knowledge on the subject in hand, while being partly the result of the half developed state of the art until very recently, is also very largely owing to narrow-mindedness on the part of some of the older members of the craft, who have largely obtained their skill by years of experience and study, some of them having expended large sums on experimental work. The same experiments have perhaps been made before, and are of necessity to be made again by others, simply because the first experimenter would not give other people the benefit of his experience. It seems at this stage in the development of refrigeration, that the improvements to be made during the next twenty years will be of very much less importance than those made during the twenty years just ending; trade secrets, so jealously guarded by some, must disappear, as they have in other branches of engineering. Storage men have been obliged to work out their own salvation in storing problems, sometimes sending their most difficult points to be answered through the columns of *Ice and Refrigeration*, and, perhaps, comparing ideas with those of their personal friends in the same line of business. It is to be observed that the most progressive and up-to-date manufacturing concerns in the United States to-day are giving their contemporaries every opportunity of observing their methods, and are very willing and anxious to talk over matters pertaining to their work, from an unselfish standpoint. So, too, the successful cold storage of the future will be sure to make " visitors welcome."

In anything which will appear in these articles, it is not the writer's intention to convey the idea that any mere theoretical knowledge, which can be acquired by reading and study, or even by an exchange of ideas in conversation, can take the place of practical observation in actual house management; but there are applications of well known natural laws, which are not generally understood by storage men, and their progress is handicapped from lack of this theoretical knowledge. The two following illustrations, bearing on temperature and ventilation, are among the common errors made in practice, yet easily understood when studied and tested: Some storage houses have formerly held their egg rooms at 33° F., fearing any nearer approach to the freezing point of water (32° F.), thinking the eggs would freeze. A simple experiment would settle this point, giving the exact freezing temperature, as well as the effect of any low temperature on the egg tissues. Again, others have thought to ventilate by opening doors during warm weather. It never happens that storage rooms can be benefited by this treatment at any time during the summer months, and only occasionally during spring and fall. The dew point of outside air is rarely below 45° F. during summer, and when cooled to the temperature of an egg room, moisture will be deposited on the goods in storage, causing a vigorous growth of mildew.

EGGS IN COLD STORAGE.

CHAPTER I.

TEMPERATURE.

TEMPERATURE is selected for first considera- Temperature more important than any other condition. tion, as it is the primary element of refrigeration, and more important than any other condition. Correct temperature alone, however, will not produce success- ful results, any more than a good air circulation, or cor- rect ventilation, would give good results with a wrong temperature. This applies more especially to egg re- frigeration, some products requiring only a low temper- ature for preservation. The common impression of cold storage is what the name implies—simply a build- ing in which the rooms may be cooled to a low degree as compared with the outside air. Even those who build, sell and erect refrigerating machinery and appa- ratus often show either gross carelessness or ignor- ance of the requirements of a house which will produce successful results. After a careful examination of some of the recently constructed houses, supposed to be strictly modern and up to date, the writer gets the impression that the architects regard temperature as the only requisite for perfect work. Some of the Unskillfully designed egg rooms. rooms in these new houses are simply insulated and fitted with brine or ammonia pipes, the location of the coils having no attention whatever, being placed, in most cases, in convenient proximity to the pipe main, and in one or two instances, the top pipe of the cooling coils was fully two feet from the ceiling. The ne- cessity of providing for air circulation seemed not worthy of consideration, to say nothing of the lack of anything like an efficient ventilating system.

Questions regarding the correct temperature of Opinions regarding correct temperature. egg rooms have been asked repeatedly of storage men who have been in the business long enough to be looked to for advice, the same person, perhaps, giving a different answer, from time to time, as his ideas changed. The query has also been asked and answered through the columns of Ice and Refrigeration a number of times. At present, however, there is no temperature on which a large majority of persons can agree as being right, and as giving superior results to any other. The claims made by the advocates of different temperatures will be considered, to determine, if possible, what degree is giving the best results in actual practice.

Questions relating to temperature.
The three questions relating to temperature were written to draw out opinion as to the right temperature, the lowest safe temperature, and what deleterious effect, if any, the egg sustained at low temperatures, which did not actually congeal the egg meat. The three temperature queries were:

First.—At what temperature do you hold your rooms for long period egg storage?

Second.—What temperature do you regard as the lowest limit at which eggs may be safely stored?

Third.—What effect have you noticed on eggs held at a lower temperature?

Figures received relating to correct temperature.
All the replies received contained answers relative to temperature, and by a very small majority 32 F. is the favorite temperature for long period egg storage. Some few, 33° F. and 34° F., with a few scattering ones up to 40° F. Under the freezing point, none recommended a temperature lower than 28° F., and for a very obvious reason, this being near to the actual freezing temperature of the albumen of a fresh egg. A very respectable minority say a temperature ranging from 30° F. to 31° F. is giving them prime results; and several recommend 30° F. straight, and say they should go no lower. In recent years there has been a decided tendency among storage men to get the temperature down near the safety limit, but many houses are so poorly equipped that

they are unable to maintain a uniform low temperature below 33° F., without danger of freezing eggs where they are exposed to the flow of cold air from coils. A house must be nicely equipped to maintain low temperatures with safety. More houses would use temperatures under 32° F., were they able to, without danger to the eggs. A very successful eastern house issued a pamphlet in 1892. At that time they maintained a temperature of 32° F. to 34° F. in their rooms. In sending out this little book during the winter of 1897–98 a postscript was added, as follows: "This pamphlet was published in 1892, when our plant was started. Since that time all first-class cold storage houses have lowered their temperatures materially." No better illustration than this can be cited to show the tendency of the times. These people now use a temperature of 30° F. for eggs.

Tendency toward lower temperature illustrated.

Most of the replies received contained answers to question No. 2, and the greater portion state this as being about 2° F. lower than that recommended for long period storage. It is presumed that these two degrees are allowed as leeway, or margin of safety, for temperature fluctuations. Some state that eggs cannot be safely held below 32° F., but give no reason why, while two or three say a temperature of 27° F. will do no harm to eggs in cases. One reply states that eggs held in cut straw at 25° F. for three months showed no bad symptoms. It has never been made clear how the package can be any protection against temperature, when the temperature has been continuously maintained for a length of time sufficient to allow the heat to escape; and we know that eggs will positively freeze at 25° F., as proven by experiments mentioned in another paragraph.

Replies to query No. 2.

The answers to question No. 3 were few in number, but cover a wide range. The scarcity of data on this point indicates that few have experimented with

Replies to query No. 3.

eggs at temperatures ranging from 25° F. to 30° F.
Some say: "Dark spot, denoting germ killed"; others,
"white gets thin"; others, "eggs will decay more
quickly"; or, "they will not 'stand up' as long when
removed from storage." It is also claimed that "yolk
is hardened or 'cooked' when temperature goes below
32° F." Some answers state a liability of freezing if
eggs are held in storage at a temperature below 32- F.
for any length of time.

Claims made by
the advocates
of high and low
temperatures. As far as possible, we will dig out reasons for the
claims made by advocates of both high and low tem-
peratures, both having equal consideration. Taking
29° F. or 30- F. and 38° F. or 40° F., as representing
the lowest and highest of general practice, we will see
what is claimed by each; and also the faults of the
other fellow's way of doing it, as they see it. Those
who are holding their egg rooms at 40° F. say it is
economical, that the eggs keep well, that the consistency of the egg meat is more nearly like that of a
fresh egg after being in storage six months, than if
held at a lower temperature. As against a low temperature they say: A temperature of 30° F. is expensive to maintain; the yolk of the egg becomes hard
and the white thin, after being in store for a long
hold; and that when the eggs are taken from storage
in warm weather it will require a longer time to get
through the sweat than if held in storage at a somewhat higher temperature, resulting in more harm to
the eggs. Some claim that the keeping qualities are
impaired by holding at a temperature as low as 30- F.,
and others note a dark spot, or clot, which forms in
the vicinity of the germ, when eggs are held below
33- F. Against this formidable array of claims, the
low temperature men have some equally strong
ones, although fewer in number. They say: There
is very much less mildew, or must, at 30° F. than at
temperatures above 32° F.; the amount of shrinkage
or evaporation from the egg is less; an egg can be

held sweet and reasonably full at this temperature
from six to eight months. This last claim is a broad
one, and very few houses are turning out eggs answering to this description.

The following, relating to high temperatures, is
quoted from a letter written by one of the best posted
men in the business, who has spent much money and
time on experiments, and studied the question for
years. He says: "A temperature of 40° F. is very
good for three months' holding, but if they run over
that, it is more than likely the eggs will commence to
cover with a white film, which grows the longer they
stand, and finally makes a musty egg." This gentleman advocates a temperature of 30° F. for long
period holding. It should be noted that the high
temperature men ignore entirely the effect of high
temperatures on the growth of this fungus, spoken of
above as a white film. The worst thing about most
storage eggs is the taste caused by this growth,
(usually called mildew or mold), which results in what
is commonly called a musty egg. To enable us to
understand the validity of these claims made by the
30° F. people, it will be necessary for us to ascertain
the conditions which are favorable, and also the conditions which are unfavorable for the propagation of
this growth of fungus, which has given storage men
so much trouble, ever since cold storage was first
used for the preservation of eggs.

Heat and moisture are the two conditions leading
to its rank growth, and the opposite—dryness and cold
—will retard or stop the growth entirely. In moist,
tropical countries many species of this parasite grow,
while in the cold, dry regions of the north its exist-
ence is limited to a single variety. The causes leading to a growth of the fungus on the outside of an egg
are not far to seek. It feeds on the moisture and
products of decomposition which are being constantly
given off by an egg, from the time it is first dropped

until its disintegration, unless immersed in a liquid, or otherwise sealed from contact with the air. This evaporation not only takes moisture from the egg, but carries with it the putrid elements from the egg tissue, resulting from a partial decomposition of the outer surface of the egg meat. Conditions of excessive moisture and the presence of decaying animal or vegetable matter, together with a moderate degree of heat, are essential to the formation of fungus of the species which are found growing on eggs in cold storage. As the heat and moisture are increased, the growth of fungus will be proportionate. Furthermore, we all understand that heat hastens decomposition, and the partial decomposition of an egg results in a growth of the fungus, as before explained, when conditions of temperature and humidity are favorable. If the temperature is low, this growth is slow; for instance, if eggs are held at a temperature of 30° F. in an atmosphere of given humidity, the growth of fungus is less rapid than if held at any temperature higher, with the same per cent of humidity. As our subject merges into humidity here, the reader is referred to what is said under this head in another chapter.

Some experiments on the freezing point of eggs.

Returning to the objections urged against low temperatures, we will see what damage is claimed from the use of a temperature of 29° to 30° F. The objections are: Liability of freezing; germ is killed; white becomes thin; yolk is hardened, and eggs will not keep as long when removed from storage. Some interesting results are obtained from experiments made by the writer. Half-rotten or "sour" eggs freeze at temperatures just a trifle under 32° F. Fresh eggs freeze at 26° to 27° F. In testing eggs which had been held in storage for several months, it was noted that the freezing point had been depressed from 1° to 2° F. An egg which is leaky will freeze at 2° to 3° higher temperature than one which is sound, probably owing to the evaporation resulting in a lower temperature. The freezing point of eggs, as above, is understood as being the degree at which they begin to form ice crystals inside. Of the replies received touching on the freezing point of eggs, nearly all agree with above experiments. The " dead germ " theory the writer has never been able to locate in fact, having never seen anything of the kind in eggs held as low as 28° to 29° F. for several weeks' time; nor in eggs held at 30° F., or a trifle under, through the season. As only two or three mention having noted this result, it would seem that some local conditions, and not low temperature, were responsible.

Dead germ theory.

The matter of the white becoming thin when eggs are held at low temperatures has some bearing; in fact, any egg held at a cold storage temperature for a long carry will show this fault, to a certain extent, especially if cooled quickly when stored, or warmed suddenly when removed from storage. With reference to the above, it is the writer's opinion that a difference of 4° to 6° F. in carrying temperature will not be noticeable in its effect on the albumen of an egg; and as to the effect of a low temperature on the egg yolk, it has been demonstrated that any temperature, which will not actually congeal the albumen, will not harm the yolk of an egg. There is a slight tendency, in this case, to a similar effect to that produced by a low temperature on cheese; that is, causes it to become " short " or crumbly.

Effect of low temperatures on eggs.

In regard to a low temperature egg not keeping as long when removed from storage, it has been the experience of the writer that no difference was noted between eggs put out from storage and the current receipts of fresh eggs, so far as any complaint or objection was concerned, the eggs being shipped in all directions, in all weathers and subject to many different conditions. A test was also made, by placing three dozen of eggs, which had been carried in storage at a temperature of 28° F. to 30° F. for five months,

A test showing keeping qualities of eggs held at low temperature.

in a case along with three dozen fresh eggs. After
three weeks no pronounced change was noted in
either, both showing considerable evaporation as a
result of exposure to the dry fall atmosphere. They
were exposed to the temperature of the receiving
room, fluctuating from 50° F. to 80° F. The eggs
from storage went through a "sweat," while the fresh
were not subjected to any such trial. As most eggs
are consumed inside of three weeks after being removed from storage, this would seem like a good
practical test of the vitality of a low temperature
egg. A mere matter of economy between holding a
room at 40° F. and 30° F., while readily appreciated
and admitted, seems of very small importance, when
a positive advantage can be obtained by carrying eggs
at the lower temperature; and a difference of 4° F. to
5° F. would be scarcely worth considering.

Low tempera-
ture prevents
"spot" rotten
eggs.

An advantage of low temperature, not yet mentioned, is the increased stiffness, or thickness, of the
white of the egg while in storage, holding the yolk in
more perfect suspension. When eggs are held at a
temperature of 36° F., or above, for any period longer
than four months, the yolk has a decided tendency to
rise and stick to the shell, causing rotten eggs, known
as "spots." It is usually understood that the yolk set-
tles; but, being of a fatty composition, it is lighter than
the albumen, and rises instead. If the albumen is
maintained in a heavy consistency, the yolk is retarded
from rising, and held in a more central position. It
was long a practice with storage men to turn eggs at
least once during the season, to prevent the above
trouble, and some recommend it even now; but the
practice has been generally abandoned with the ad-
vent of low temperatures for egg storing.

Putting eggs in
and removing
from storage.

When eggs are put in cold storage they should not
be cooled rapidly. The effect on the egg tissues is
bad—they should have time to rearrange themselves
to the changed temperature. This is especially true

where eggs are placed in storage in extreme warm weather. Sudden warming is also detrimental to the welfare of an egg, for a similar reason to above. The most noticeable effect of either is a thinned albumen. If this process of cooling and warming could be practiced carefully (which is not always practicable commercially), a well kept storage egg would come out of storage with nearly the same vitality it had when fresh.

CHAPTER II.

HUMIDITY.

INFORMATION on the subject of humidity, as applied to the cold storage of eggs, is very meager. Not more than a dozen of the replies received in answer to the list of inquiries sent out contain information on the three queries under the head of humidity. Considering the amount of talk we have all heard, with dry air as a subject, this scarcity of knowledge is rather surprising. Those who have had experience with cold storage work and the products handled are well aware that an essential for good results in egg refrigeration is a dry atmosphere in the egg room; but just how dry, very few are able to give even an approximate estimate. Very likely if a cold storage man is asked in regard to it, he will reply that an egg room should be "neither too moist nor too dry." What this "happy medium" is, that will not shrink or evaporate the eggs badly, and yet keep down the growth of fungus to a minimum, is what all are striving for, and very few have the means of

knowing when this point is reached. A few years ago a prominent commission man, in conversation with the writer, speaking of storage eggs, said: "You storage men are between the devil and the deep sea. You always shrink 'em or stink 'em"; meaning that eggs which were held long in storage would show either a considerable evaporation or a radical "musty" flavor. To some extent this is true, but with a penetrating circulation, careful ventilation and a judicious use of absorbents (all of which will be considered under their proper heads) egg can be, and are, turned out of storage without this strong, foreign flavor, and with little evaporation or shrinkage.

The questions relating to humidity were written with a full understanding of the scarcity of information on the subject, and were designed to locate, if

18

possible, those who were making tests of air moisture, and get opinions on the correct humidity for a given temperature. The following are the queries:

First.—What tests, if any, have you made of the dryness or humidity of your egg rooms?

Second.—What per cent of air moisture do you find gives the best results at the temperature you use?

Third.—What instrument do you use for testing air moisture?

Questions 1 and 3 are practically the same, the latter being written simply to make the query more plain and indicate whether an instrument or some other test was used for determining air moisture. Four houses reporting are using the dry and wet bulb thermometers; the others are using hygrometers of French or German make.

The answers to question 2 vary greatly; some also giving the actual testing humidity of their rooms and their opinion of a correct degree as well. From 70 to 80 per cent of humidity is the test of nearly all reporting, and of the rooms tested by the writer, nearly all show a similar humidity, with one occasionally going as high as 85 per cent, and some as low as 65 per cent. Two answers recommend a humidity of 65 per cent, and one a humidity of 60 per cent, with a temperature of 30° F. to 32° F. Others hold that their testing humidity of 70 to 80 per cent is correct. The matter of correct humidity will be discussed further on. *Replies relating to correct humidity.*

The humidity of a room depends on the season to a moderate extent, and the condition of the room, as regards ventilation, in some cases. In late fall or winter, especially, if air is taken directly into the room from the outside, the humidity will be low. As cool weather approaches, the tendency is for the humidity to rise, and unless kept down by ventilation or by the use of absorbents, serious consequences are sure to follow. *Influences controlling humidity.*

What relative humidity signifies.

To enable us to thoroughly understand the meaning of relative humidity, as it is called, we will study a few extracts from " Instructions to Voluntary Observers," issued by the Weather Bureau at Washington, D. C. Humidity is considered on a decimal scale, with 100 the saturation point of the air, at which it will hold no more water vapor, and 0 the point at which air contains no moisture whatever. The various percentages between these points is a degree of humidity relative to these two extremes, or relative humidity. The quotations below are not contained in the recent issue of instructions, but are from the issue of 1892, which is now superseded by that of 1897.

WATER VAPOR IN AIR.

Quoted from "Instructions to Weather Observers."

The air contains vapor of water, transparent and colorless like its other gaseous components. It only becomes visible on condensing to fog or cloud, which is only water in a fine state of division. The amount is very variable at different times, even in the vicinity of the ocean. The amount of moisture that can exist as vapor in the air depends on the temperature. There is a certain pressure of vapor, corresponding to every temperature, which cannot be exceeded : beyond this there is condensation. This temperature is called the temperature of saturation for the pressure. When the temperature of the air diminishes until the saturation temperature for the vapor contained is reached, any further fall causes a condensation of moisture. The temperature at which this occurs is called the dew point temperature of the air at that time. The less the quantity of moisture the air contains, the lower will be the temperature of the dew point. For different saturation temperatures, the weight of vapor, in grains, contained in a cubic foot of air is as follows:

Temperature of Saturation, Degrees F.	Weight in a Cubic Foot, Grains.
0	0.56
10	0.87
20	1.32
30	1.96
40	2.85
50	4.08
60	5.74
70	7.98
80	10.93
90	14.79
100	19.77

The air is never perfectly saturated, not even when rain is falling; neither is it ever perfectly dry at any place. Relative humidity expresses relative amount of moisture in the air only as long as the temperature of the air remains constant. For this reason relative humidity is an imperfect datum. At a low temperature, even a high relative humidity represents a very small amount of vapor actually in the air, while a low relative humidity at a high temperature represents a great deal.

The most important law relating to above concise statements, and one which, if carefully noted and applied, will make all work in humidity easily understood, is best expressed thus: *The capacity of air for moisture is increased with its temperature.** Law governing air moisture.

At a temperature of 40° F., air will hold in suspension more water vapor than at any lower temperature (see table); and when the difference is as much as 10° F., the difference in the amount of moisture the air will hold is very considerable. To illustrate: Air which is saturated with moisture at 30° F., when raised in temperature to 40° F., then holds but 68 per cent of its total capacity.

Under the head of "Temperature," it is stated that: "If eggs are held at a temperature of 30° F. in an atmosphere of a given humidity, the growth of fungus is less rapid than if held at any temperature higher, with the same per cent of humidity. Referring again to the table, we see that a cubic foot of air, when saturated at a temperature of 40° F., contains 2.85 grains of water vapor, while at 30° F. it contains but 1.96 grains, or only about two-thirds as much as at 40° F. The same holds true with any relative humidity, the same as when the air is saturated. Take, for instance, air at a temperature of 40° F., with a humidity of 75 per cent, then a cubic foot of air holds 2.14 grains of water vapor per cubic foot; and at a temperature of Variation of humidity with temperature illustrated.

---

* Strictly speaking, air has no capacity for moisture, the water vapor being simply diffused through the air, after the nature of a mechanical mixture. For all practical purposes, we may regard it as being absorbed by the air, and it is usually so treated.

30° F., with the same relative humidity, it would hold
but 1.47 grains. This great difference in the amount
of moisture contained in the air at different temperatures, and still having the same relative humidity,
has as radical an effect on the growth of fungus
as does the difference in temperature. This is no
mere theory, as the writer has demonstrated it, to his
own satisfaction, at least, during several seasons'
observation. If it is hoped to keep down the growth
of fungus in a temperature of 40° F. by maintaining an
atmosphere with a lower relative humidity, the result
is a badly evaporated egg, which loses its vitality and
value very rapidly when held in storage for a term
exceeding three or four months; the white becomes
thin and watery, with a strong tendency to develop
"spot" rotten eggs. As the fullness or absence of
evaporation is of only secondary consideration to their
sweetness, when eggs are tested by buyers, it is
necessary to prevent this trouble if the eggs turned
out from storage are to be considered first-class.

Result of too
dry an egg
room.

From the foregoing it seems clear that to turn out
sweet eggs at a temperature of 40° F. it is necessary
to maintain a lower relative humidity than at any temperature lower, and the result cannot fail to be as described. The writer has already given a summary of
the replies to the questions relating to humidity, which
are few in number, and not very complete. A little
is better than nothing, however, and by comparing
his own data with the results obtained by others, and
paying careful attention to their opinions, the following table of correct humidity for a given temperature
in egg rooms has been compiled. There are no data on
the subject in print, so far as known, and no claim for
absolute accuracy is made in presenting this first
effort in that direction, but as the figures are taken
from actual results, no great mistake can be made by
depending on them. The percentages of humidity
given are modified, to some extent, by the intensity

Correct relative
humidity not
accurately
known.

and distribution of the air circulation employed. (See Chapter III on "Circulation.")

CORRECT RELATIVE HUMIDITY FOR A GIVEN TEMPERATURE IN EGG ROOMS.

Temperature in Degrees F.	Relative Humidity, Per Cent.	Humidity table for egg rooms.
28	80	
29	78	
30	76	
31	74	
32	71	
33	69	
34	67	
35	65	
36	62	
37	60	
38	58	
39	56	
40	53	

There are two kinds of instruments in use for de- Hygrometers. termining humidity, hygrometers and psychrometers. The hygrometer depends on the expansion and contraction of some substance, as a human hair, in the presence of more or less moisture in the air. The hair used is fastened at one end, the other end passing around a pulley, to which is fastened a pointer, which moves over a graduated arc as the hair changes its length. The scale reads from 0 to 100. The chief advantage of these instruments is that results are obtained at once, the reading corresponding to the percentage of saturation or relative humidity; but these instruments are affected by changes of temperature, and shocks or vibration materially affect the reading. Further, they are more expensive in first cost, and not so convenient to use, as they must hang for some time in the room to be tested, while with the sling psychrometer, described in another paragraph, an observer can pass from room to room, getting observation in less than two minutes in each room, needing but one instrument and making all observations at practically the same time.

A psychrometer is simply two thermometers Psychrometers. mounted on a frame; the bulb of one being covered with muslin so as to retain a film of water surrounding it. The working of this instrument depends on a law which may be roughly expressed, as "evaporation carries off heat." The evaporation of water from the bulb incased in muslin, known as the wet bulb, cools it somewhat, depending on how dry the air surrounding it may be. The difference between the reading of the wet bulb thermometer and the reading of the dry bulb thermometer, when compared with reference to a prepared table, gives the relative humidity of the air at the time of making the observation. Psychrometers are of two kinds, stationary and sling.

Stationary psychrometer. The stationary psychrometer is essentially like the sling psychrometer, both depending on the same principle. The sling instrument is more compact and provided with a handle for whirling, while the stationary instrument is intended to be fastened against the wall, or on a post, the muslin covering the wet bulb being connected by a porous cord with a reservoir of water, to keep the supply of water continuous. This is essential, as it takes some little time to obtain a correct reading with this pattern of instrument. For this reason it is open to the same objections as the hygrometer. Also, after short use the muslin covering the wet bulb, and the cord feeding water to it, become clogged with solid matter and fungous growth affecting its accuracy. At any temperature below 32° F. this instrument is useless, as the water will freeze in the cord supplying the muslin on the wet bulb, and the muslin becomes dry in consequence.

Sling psychrometer. For practical, accurate and quick results at any temperature there is no instrument so reliable and convenient as the sling psychrometer, preferably of the pattern known as Prof. Marvin's improved psychrometer, shown in the illustration. This is a standard Weather Bureau instrument, and when used

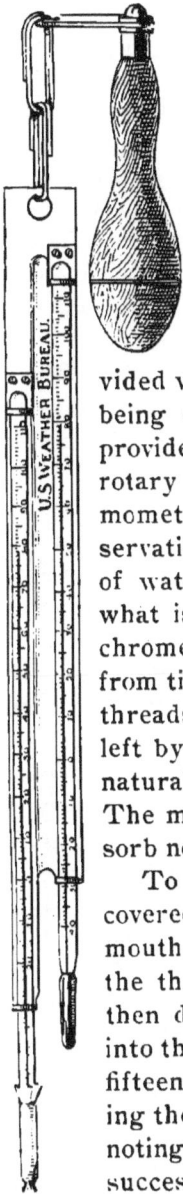

in connection with the tables of humidity published by the bureau, any needed results may be obtained with a fair degree of accuracy. The sling psychrometer, as illustrated, consists of a pair of thermometers mounted on an aluminum plate, one higher than the other, the lower having its bulb covered with a small sack of muslin. At the top, the frame or plate supporting the thermometers is provided with a handle for whirling, this handle being connected by links to the plate, and provided with a swivel to allow of a smooth rotary motion. The bulb of the lower thermometer is wet at the time of making an observation, the muslin serving to retain a film of water, surrounding and in contact with what is known as the wet bulb of the psychrometer. The muslin should be renewed from time to time, as the meshes between the threads will gradually fill with solid matter left by the evaporation of the water and the natural accumulation of dust from the air. The muslin in this condition will neither absorb nor evaporate the water readily.

To make an observation dip the muslin covered bulb in a small cup or other wide-mouthed receptacle containing water. Whirl the thermometer for ten or fifteen seconds, then dip the wet bulb of the psychrometer into the water again. Whirl again for ten or fifteen seconds, stop and read quickly, reading the wet bulb first. Repeat once or twice, noting the reading each time. When two successive readings of the wet bulb agree very nearly, the lowest point has been reached. Dip the wet bulb only after the

Directions for using the sling psychrometer.

SLING PSYCHROMETER.

first whirling, as this is done only to make sure
that the muslin is thoroughly saturated with water.
If the water used is of nearly the same temperature
as the room, correct readings are sooner obtained.
If the psychrometer and water are at a much higher
temperature than the air of the room, it will take a
proportionately longer time to reach a correct reading, but the accuracy will not be impaired, if sufficient
time is allowed for the mercury to settle. It is very
important that the muslin covered bulb should not
become dry in the least; it should be saturated with
water during the full time of observation. There will
be no difficulty in getting accurate readings down to
29° F., as indicated by the dry bulb. At about this
temperature, and with the wet bulb at about 27° F.,
ice will form on the wet bulb and cause the psychrometer to become somewhat erratic in its behavior.*

Whirling and
stopping sling
psychrometer.

It is difficult to describe the proper movements
for whirling the sling psychrometer, a little practice
being the best instructor. The handle is held in
a horizontal position, the frame mounting the thermometers revolving around the pivot, after the manner of the weapon with which David slew Goliath, and
from which our moisture-tester gets the easy part
of its name. A high rate of speed is unnecessary, a
natural, easy motion of the forearm or wrist being all
that is required. When stopping the psychrometer
the arm should follow the thermometer from the highest point of the circle of rotation, whereby the radius
of the path of the psychrometer is increased, and the
momentum overcome. The stopping can be accom-
plished in a single revolution, after a little practice.
The psychrometer will come to rest very nicely by
simply allowing the arm to stand still, but the final
revolution will be quite irregular and jerky.

* The writer is in receipt of a special report on this point, prepared by
Prof. C. F. Marvin, in charge of the instrument division of the Weather Bureau,
and will gladly give any one having difficulty with the psychrometer at these
temperatures information so far as he can; but the point involved is somewhat
intricate, and so few are using temperatures as low as 29° F. that it is thought
best to omit a discussion of this phenomenon.

In making observation in a storage room, the psy- Making obser-
vations and
chrometer should be held as far as convenient from reading sling
psychrometer.
the body, and toward the direction from which the
circulation comes—the observer standing to the lee-
ward, as it were. In some cases it is necessary, or
advisable, to step slowly back and forth a few steps,
and the observer should turn his head from the direction of the psychrometer, so his breath will not
affect the reading. In reading a thermometer, read
as quickly as possible, and do not allow the breath to
strike the bulb. It is a common practice with the
writer to hold his breath while reading a thermometer. It is unnecessary to caution against allowing the
psychrometer to strike any object while whirling.
In case it should, the observer will have $5 worth of
experience, but no psychrometer.

RELATIVE HUMIDITY, PER CENT.

(Dry ther.)	Difference between the dry and wet thermometers $(t-t')$.												(Dry ther.)	Table of rela- tive humidity.
	0°.5	1°.0	1°.5	2°.0	2°.5	3°.0	3°.5	4°.0	4°.5	5°.0	5°.5	6°.0		
28	94	88	82	77	71	65	60	54	49	43	38	33	28	
29	94	89	83	77	72	66	61	56	50	45	40	35	29	
30	94	89	84	78	73	67	62	57	52	47	41	36	30	
31	95	89	84	79	74	68	63	58	53	48	43	38	31	
32	95	90	84	79	74	69	64	59	54	50	45	40	32	
33	95	90	85	80	75	70	65	60	56	51	47	42	33	
34	95	91	86	81	75	72	67	62	57	53	48	44	34	
35	95	91	86	82	76	73	69	65	59	54	50	45	35	
36	96	91	86	82	77	73	70	66	61	56	51	47	36	
37	96	91	87	82	78	74	70	66	62	57	52	48	37	
38	96	92	87	83	79	75	71	67	63	58	54	50	38	
39	96	92	88	83	79	75	72	68	63	59	55	52	39	
40	96	92	88	84	80	76	72	68	64	60	56	53	40	

The above short table needs no explanation fur- Weather
Bureau table.
ther than has been already given. It will cover any
case in egg room observations. This table was not
intended for cold storage work, being a part of the
regular humidity tables published by the Weather
Bureau. The full set of tables can be had by

addressing the chief of the Weather Bureau, Department of Agriculture, Washington, D. C. They are published in pamphlet form, along with tables giving dew point temperatures. Observers must work out the small fractions for themselves, if they think necessary, but results within the limits covered by the table are near enough for present practical purposes.

It is of no use to test for moisture unless having the ability to control it, any more than a thermometer would be of use unless the means of regulating temperature were at hand. Humidity can be controlled by ventilation and the use of absorbents, which are considered elsewhere.

CHAPTER III.

CIRCULATION.

A vigorous and penetrating circulation of air must be maintained in a cold storage room for eggs if good results are to be insured. The importance of this condition, as applied to eggs especially, is quite generally appreciated, and it is noticeable that the warehouses producing the most perfect work have scientific and carefully designed air circulating systems. It is also a fact that a strong, searching circulation will do much to counteract defects in a cooling apparatus, or wrong conditions in the egg room in some other particular. In proof of this, the writer is familiar with a number of successful houses where prominence is made of the air circulating system only, some of the other conditions being neglected altogether, or attended to in a perfunctory manner.

Before going farther, it is best that we separate circulation from its tangle with ventilation. These two terms are quite commonly confused when applied to cold storage work. Circulation, as here discussed, applies to the motion of air within the storage room— air currents resulting from a difference in temperature of the air in different parts of the room, or the result of mechanical force applied to the air by use of fans, blowers or exhausters. In distinction from circulation, ventilation means the renewal of the air of a storage room, either by forcing fresh air from the outside atmosphere into the storage room, or by exhausting the foul air from the room. Ventilation is not under consideration here, but will be taken up as a separate subject.

The reason why a vigorous and well distributed circulation of air in an egg room will give superior results over a sluggish or partial circulation may not be readily apparent. A circulation of air is of benefit in combination with moisture absorbing capacity in

29

the form of frozen surfaces or deliquescent chemicals. Stirring up the air merely, as with an electric motor fan, without provision for extracting the moisture, is of doubtful utility, and may, in some instances, prove positively detrimental, as it is liable to cause condensation of moisture on the goods, or walls of storage room, instead of its correct resting place: the cooling coils and absorbents. Let us see how the circulation of air in a storage room operates to benefit its condition.

How air is purified by circulation. Under head of temperature, we have seen that the evaporation from an egg contains the putrid elements resulting from a partial decomposition of the egg tissues, and that the air of a storage room carries them in suspension. It is probable that these foul elements are partly in the form of gases absorbed in the moisture thrown off from the egg; and if, therefore, this moisture is promptly frozen on the cooling pipes, or absorbed by chemicals, the poisonous gases and products of decomposition are very largely rendered harmless. This is also true of the germs which produce mold and hasten decay, which are ever present in the atmosphere of a storage room, being carried to a considerable extent by the water vapor in the air, along with the foul matter of various kinds referred to. If the vapor laden air surrounding an egg is not removed and fresh air supplied in its place, the air in the immediate vicinity of the egg gets fully charged with elements which will produce a growth of fungus on its exterior, affecting and flavoring the interior—the flavor varying in intensity, depending on how thoroughly impregnated with fungus-producing vapor the air in which the egg is kept may be. In short, then, circulation is of value because it assists in purifying the air. It should be kept up so that the air may be constantly undergoing a purifying process to free it from the effluvia which are always being thrown off by the eggs, even at very low temperatures. It has been suggested that a brisk circulation of air which will keep in motion the whole volume of air in the storage room will have a purifying influence independent of any moisture absorbing capacity, but no satisfactory reason has been assigned. There may be such an influence operative when the air is mechanically circulated. If so, there seems to be no scientific or practical explanation of it.

Many patents have been granted for improvements in storage rooms or refrigerators using ice only for a cooling agent; house refrigerators, refrigerator cars and refrigerator buildings are represented in number about in order named. A large portion of the patents granted have been on claims for the improvement of circulation, and this is the keynote of whatever success has been attained by the various systems which use ice only for cooling. As any system of cooling whereby the air is caused to circulate in contact with melting ice is now quite generally regarded as obsolete for the successful refrigeration of eggs for long period storage, a discussion of the merits of the various devices applied to this work will be omitted.* It may be said to their credit, however, that the builders of ice refrigerators have originated ideas on circulation which have been of much value to the present-day refrigerating engineers, and there are still those who may obtain good information from this source—the provision for circulation (or rather, lack of circulation) in a few of the new mechanical cold storage houses being simply ridiculous.

In the more progressive mechanically refrigerated houses there are a number of devices, which have been introduced for assisting natural gravity air circulation, also the various modifications of the me-

Circulation in natural ice refrigerators.

Various methods in use for promoting circulation.

---

* The above must not be construed as condemning the use of ice as a refrigerant when rightly applied. The writer has in successful operation a system of gravity brine circulation, cooled by ice and salt, with which he will undertake to produce at moderate expense any possible results in refrigeration down to a temperature of 15° F.

chanical forced circulation system. Some of these will continue to gain favor because of the improved results obtained by their use. The main requisite in any air circulating system is an ability to cause an equal distribution of the moving air, as it comes from the moisture absorbing surfaces, forcing it uniformly to all parts of the room and compelling it to flow through and around the, piles of stored goods. As a secondary consideration may be mentioned the equalization of temperature and humidity in all parts of the room. The writer is somewhat biased in favor of forced circulation, having developed a very complete system on this line, with some new features. Aside from a matter of economy of space and operation, the system employed matters little, if an effective circulation is produced.

Questions relating to circulation.

The questions bearing upon circulation contained in the list of inquiries sent out by the writer are as follows:

First.—In piping your rooms what provision was made for air circulation?

Second.—What difference in temperature do you notice in different parts of the same room?

Third.—Do you use a fan or any kind of mechanical device for maintaining a circulation of air in the rooms?

Answers to Query 1.

More answers were received on this subject than on the subject of humidity, but not exceeding one-third contained tangible replies to all three inquiries. Several of the answers confounded circulation with ventilation, as before alluded to. Question 1, in particular, was badly neglected, indicating, no doubt, that no provision was made for circulation in a majority of cases. The common device in use for causing air to circulate more rapidly over the cooling coils, when they are placed directly in the room, is some form of screen, mantle, apron, false ceiling or partition, as illustrated in Figs. 3, 4 and 5. Many of these have been put up after the house has been in operation for

some time, and are very crude affairs, applied in all
conceivable combinations with the pipe coils. In some
cases canvas curtains, or a thin wooden screen, have
been suspended under ceiling coils with a slant to
cause the cold air to flow off one side, and with surprising improvement to the room, considering the
simplicity of the device. Forced circulation with a
complete system of distributing air ducts is coming
into general use, as the merits of this way of producing
circulation are better understood and appreciated.

Query 2 was answered more generally, but that Answers to
some of the answers were mere guesses, or state- Query 2.
ments made without testing, is very evident, as they
state that no difference was noticed in different parts
of the same room. With open piping or gravity air
circulation, this is an impossibility—it is only possible
with a perfectly designed forced circulation system.
In contrast to this claim some answers state a difference in temperature of as high as 4° F. to 5° F., but
most answers show a difference of 1° F. to 2° F.; a
few ½° F. to 1° F.; and, still others, as before stated,
none at all. A marked variation of temperature in
different parts of a room, while in most cases caused
by defective circulation, is due sometimes partly to
location of room as to outside exposure, proximity to
freezing rooms, character of the insulating walls, etc.
An egg room placed over a low temperature freezing
room will show more variation between floor and
ceiling than when located over another egg room,
conditions being otherwise the same. Where this
arrangement occurs, and the egg rooms are operated
on a natural gravity air circulation system, eggs may
be frozen near the floor, when a thermometer hanging
at the height of a person's eyes would read 30° F. or
above. Even with the very best insulation, the result
of this very common arrangement is a defective circulation and more or less variation in temperature
between floor and ceiling.

In reply to Question 3, about a dozen state that they are using some form of mechanical forced circulation. The advantages of this method will be discussed quite fully later on. About double this number are using the small electric fans. These also will be treated in the discussion of mechanical air circulation in another chapter.

As air circulation is a somewhat neglected subject, and comparatively few have experimented enough to have positive opinions, based upon practical experience, regarding the merits of different devices and methods, some of the more prominent and successful ones are illustrated and discussed in this article.

In considering the following outlined arrangements of piping in the storage room and the various locations of screens, partitions, etc., in combination with the coils, for the purpose of separating the warm and cold currents of air (the one on its way upward from lower part of room to the top of cooling coils; the other downward from cooling coil toward floor), the principle on which this movement of air operates should be noted, so that the underlying law may be understood. The cause of a circulation of air in a storage room with direct piping is a variation of temperature, which causes a difference in weight of the air in different portions of the room. The air in the immediate vicinity of the pipes is cooled to a lower temperature than in the balance of the room, causing it to drop toward the floor by reason of its greater specific gravity—what is designated as gravity air circulation. Just as long as the flow of the refrigerant is maintained within the cooling pipes, the air will circulate by the action of gravity, the lighter warm air in top of room descending to replace the air in contact with pipes, which falls, as cooled, toward the floor. Should the refrigerant passing through coils be shut off, the cooling effect is checked, and as a result air circula-

tion over the pipes ceases. This should make plain
the fact that uniform temperatures in all parts of the
room are not even an approximate possibility in any
room depending on natural gravity air circulation. It
may also be observed that the eggs exposed to the
flow of cold air near bottom of coils will stand in a
dryer and colder atmosphere than those in top and
center of room.

Fig. 1 shows an outline sketch of piping suspended Open overhead
piping.
from the ceiling of a room—the most unscientific way
possible for a room to be piped, as it provides for no

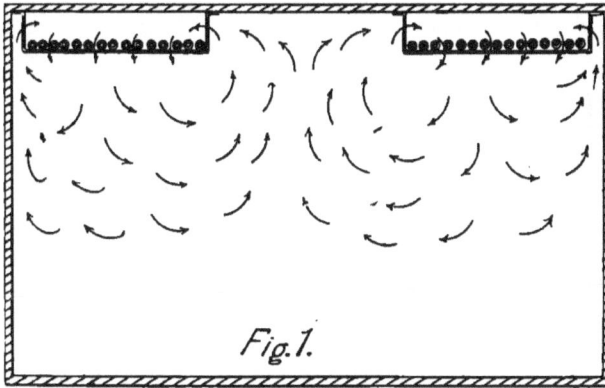

Fig. 1.

air circulation whatever. The only possible reason
why air will circulate over pipes in this position is because of the fact that the whole ceiling is not covered
by pipes, which allows of a partial circulation, as
shown by the arrows. The volume of circulation in
the lower half of room is practically nothing. It is
largely confined to the top of room, the lower part
being cooled by conduction and radiation almost entirely. It may be asked: How can a room be cooled
by radiation? In the same way that a room is heated
by radiation, except that in cooling a room the heat is
radiated *from* the objects in the room, and not *to* them,
as when heating. This gives us ample reason why a
room should be cooled by circulating the air over

frozen surfaces located outside of the room, or at
least in a position so that no radiation or conduction
can occur. The use of insulated screens or mantles,
as shown in Figs. 3, 4 and 5, is recommended as being
superior to any arrangement of open piping; but, of
course, it is not equal to forced circulation, in which
the pipes are located outside of room entirely.

Open side wall
piping.

Fig. 2 shows another very common and faulty
arrangement of piping for cooling an egg room. The
only improvement over the arrangement shown in
Fig. 1 is that it allows of a moderate action of gravity

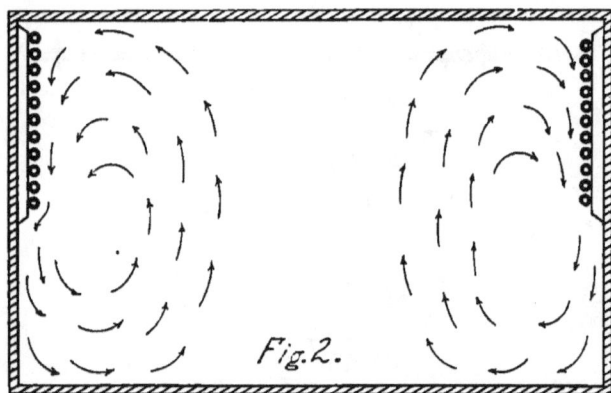

Fig. 2.

near the coils, as shown by the arrows. It is open to
the same objection on the ground of conduction and
radiation as No. 1, but to a lesser degree. The coils
are placed a few inches out from the walls, to allow
the air to circulate around the pipes freely, and to
provide room for an accumulation of frost. The top
of the coil should be quite close to the ceiling. If the
coil is placed, say midway between floor and ceiling
(unless it covers nearly the whole space), it is sure to
result in the air becoming stratified, a warm layer of
air in top of room resting on a colder one near floor,
perhaps to an extent so great as to cause a difference
of 10° F. in temperature between floor and ceiling of
room. A case with exactly these conditions is on

record. Another very bad arrangement of side wall
piping came to the notice of the writer recently. A
room exceeding fifty feet square was piped completely
around from floor to ceiling with the exception of the
doors. Circulation could penetrate but a comparatively small portion of the space in this room, and in
a large area of the central portion the air was consequently very foul, and mold and must were rampant.

Fig. 3 gives us the first primitive improvement Apron and
side wall
over open coils, and it is a long step in the right direc- piping.
tion, but it fails to take care of the center of the room,

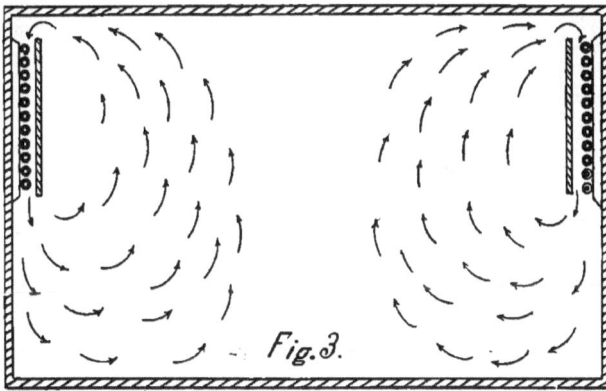

Fig. 3.

especially near the ceiling. The usefulness of this
device consists in its ability to increase the velocity,
and consequently the volume of air passing over the
cooling coils. The increased velocity of air causes it
to cover a greater area, and spread toward the center
of room further. The apron or screen used before
the coils should be constructed of any moderately
good non-conductor. Separating the warm and cold
currents of air increases the draft, on the same principle that a fire burning in a flue creates a greater
suction or a more rapid displacement of air than when
burning in the open.

Fig. 4 is simply an addition to No. 3, of a false False ceiling
and apron ap-
ceiling or curtain extending well out toward the plied to side
wall piping.

center of the room. This obliges the circulation to
spread so as to cover a large portion of the cross-section area of the room, as indicated by the arrows, but
has the effect of reducing its volume to some extent.
This ceiling apron should have a slant of not less than
one foot in ten. It occupies some considerable space,
but is richly worth it. The opening into outer edge
of apron in center of room need not exceed three
inches in depth in most cases, and, as some space
must be left at the top of room for air circulation

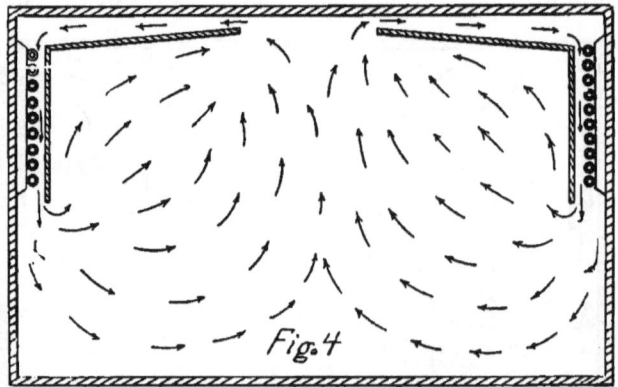

Fig. 4

with the wall coils, without ceiling apron, not much
space is wasted by its addition.

Gay's system
of box coils.   Fig. 5 gives us an entirely different arrangement
of piping, but with essentially the arrangement of
aprons shown in Fig. 4. This is the system advo-
cated by Mr. C. M. Gay on page 106 of the August,
1897, number of *Ice and Refrigeration*, and the
writer believes it to be the best idea for air circulation of any having pipes directly in the room. The
following is quoted from Mr. Gay's description:
"Upper pipes of box coils should be about ten inches
below the ceiling of the room, to prevent sweating.
When brine or ammonia is turned into these pipes
(as shown in Fig. 5), the cold air around the pipes
seeks an outlet downward and passes between the

false partition and the side wall of the room, thus
displacing or pushing along the air in center of room,
the cold air naturally seeking the lowest point and the
warm air the highest point, each by reason of its relative gravity. Thus as the cold air falls from the cooling surfaces it is replaced by the warm air from
highest point in center of room. This secures a
natural circulation and a dry room, there being no
counter-currents nor tendency to precipitate moisture
on walls or ceiling."

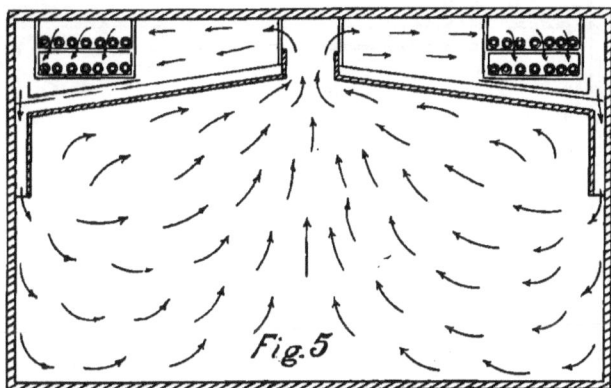

Fig. 5

Fig. 6 is the St. Clair or pipe loft system, which St. Clair, or
has been applied to many remodeled overhead ice pipe loft
system.
cold storages, by placing the pipes in a part of what
formerly was the ice space, and, in some cases, using
the original air ducts for circulation. The sketch
here shown represents one room only, but as many as
five or six different floors have been operated from a
single pipe loft, using one main air duct for the down,
and one for return air circulation, each floor having a
connection with the mains in which the flow of air is
regulated by gates. A better arrangement, when
more than one floor is to be operated on this system,
is to have independent ducts for each room, and the
cooling coils separated likewise; then any room or
rooms may be used for other products at any time

when free of eggs. This latter arrangement, of course, requires more space and is slightly objectionable on this account. The circulation is more vigorous with this system than with any pipe-in-the-room system, depending on the law that the higher the column of air the stronger the draft, on the same principle that a tall chimney gives greater draft than a short one. It is, therefore, better than any room piping, and has the added advantage of

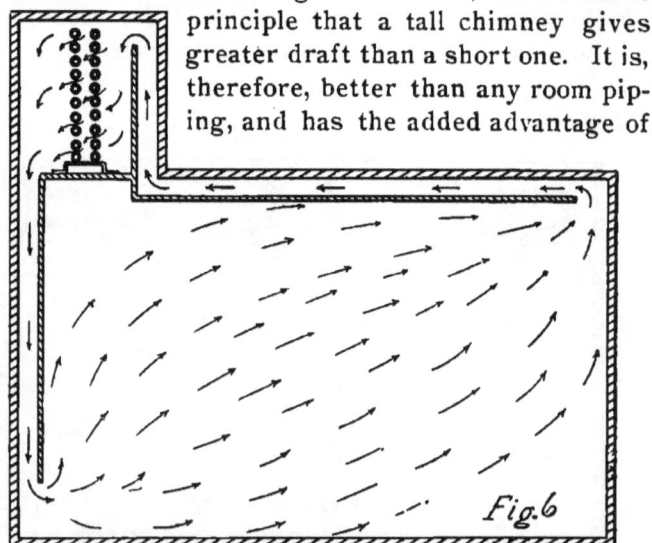

Fig.6

being easily shut off from the room, when the weather no longer requires cooling power. The need of keeping the air of the room from contact with the frost on pipes will be looked into under ventilation and absorbents.

Size of egg rooms.

Refrigerator rooms for the storage of eggs should not exceed thirty or thirty-five feet in width. The cross-section illustrations of rooms cooled by gravity air circulation, which have already been illustrated, and the two sketches shown herewith of the arrangement of air distributing ducts used in two systems of forced circulation, are sufficient to show why a room should not be excessively wide as compared to its height. In a wide room it may be seen that the air from cold air ducts, in case of forced circulation, or from the bottom of cooling coils in case of room piping, is required to pass over more eggs in its flow to the return air duct or false ceiling. The eggs, then, are not all exposed to the same drying and purifying influence, because the air as it comes from the cooling coils is at its maximum dryness and purity, and becomes impregnated with moisture and impurities more and more as it flows through the goods. If the length of piles of goods is great from side to center of room, the eggs in top and center of room will be exposed to air which is much more impure and moist than the eggs first exposed to the flow of air directly from cooling coils. This applies more especially to the gravity systems of air circulation. With forced circulation, the air moves probably three or four times as fast as when a gravity system is used, consequently the air in top and center of room does not carry the amount of impurities that it does if depending only on gravity for its motion. This fact in itself is a very good reason why forced circulation is superior to any gravity system.

It has been claimed that eggs will lose weight by shrinkage more rapidly when stored in a room in which the air is circulated by mechanical means than in a room operated on the gravity air circulation principle. This assertion is based on the assumption that the air is circulated at a much higher velocity when forced circulation is employed, and is only partly true because no account is taken of humidity. If the humidity was the same in both cases, the claim would be strictly true. Every intelligent housewife knows that linen hung in the open air to dry, will be freed of moisture quicker when a strong breeze is blowing, than when the air is nearly still. The same principle applies moderately to eggs in a refrigerator room. With the same per cent of moisture, the more rapid the circulation the greater the evaporation from the eggs; but if the facts were known, it would be

Relation between circulation and humidity.

found in every instance where trouble from excessive shrinkage of the egg meat was experienced, that no attempt was made to regulate humidity. It is as easy to control humidity as it is to control temperature, and with no bad effect on the other conditions in the storage room, if we go about it in the right way. Ventilation and the use of absorbents are agents which can be utilized for this purpose.

Proportioning humidity to the circulation.   With a vigorous circulation of air, an egg room may be maintained at a humidity which would be disastrous, if only a sluggish circulation was operative. Why? Because a brisk movement of air around the eggs removes the moisture and impurities as fast as given off by the eggs. They are not allowed to remain in the vicinity of the eggs to work mischief, but are promptly hurried off to the cooling coils or absorbents, where they are, for the most part, rendered harmless. This seems to explain how eggs may be carried sweet, and with very little evaporation, when a well distributed forced circulation of air is employed. With any gravity system, the circulation of air cannot be controlled, because depending on the temperature of refrigerant flowing in the pipes for its velocity; and as the temperature of refrigerant is regulated to correspond with outside weather conditions (lower in warm weather and higher in cold weather) the velocity of circulation is not constant—being least in the cold weather of fall, when it is most needed. With a good system of forced circulation installed, the problem, then, is to proportion the circulation of air to the humidity. We might take our humidity at the degree which would come naturally, if no attempt were made to control it, and speed our blower up or down to produce a circulation to match, but it would probably be best to provide a circulation which would handle a large volume of air at a brisk speed, and raise our humidity to as high a point as would be safe. Referring to the table of correct humidity given in Chapter

II, page 23, it will be noted that an arbitrary percentage is given for each temperature. These are the most desirable percentages of air moisture for average conditions of circulation, as when using the most scientific forms of gravity air circulation during warm weather. When using a good system of forced circulation, these percentages may be increased moderately. Exactly how much will depend on conditions, and can be told only by trial, but it will be much greater in proportion at the high temperatures, ranging from 2 per cent or 3 per cent on the low, to 7 per cent or 8 per cent on higher temperatures.

Before taking up the forced circulation systems proper, the electric fans used in so many large houses will be considered. These little fans are a four to six-bladed disk fan, from twelve to eighteen inches in diameter, attached directly to the shaft of a ⅛ or ¼-horse power electric motor. The wires supplying the electric current to the motor are usually connected to the socket for an ordinary incandescent lamp. Electric fans are usually placed on the floor in the back end of alleyway, or in an opening in the piles of goods, creating a draft of air from one extremity of the room toward the other. As the air from the fan will follow a path of least resistance, the circulation resulting from their operation is largely confined to the alleyways or openings in the piles of stored goods—it does not penetrate through and behind the piles of eggs. It may be readily observed that this is of doubtful utility, and may at times lead to positive harm by causing a condensation of moisture upon goods as a result of the comparatively warm, moisture laden upper strata of air coming in contact with the flow of cold air from the cooling pipes. Electric fans have also been utilized to propel the air from the cooling pipes, for which purpose they are placed in an opening in a screen or mantle covering the pipes, forcing the cooled air outwardly into the

Electric fan in egg room not desirable.

room. In any other position, they are only useful as a "talking point," as it is likely to impress a prospective customer favorably with the cooling power of a refrigerator, to allow him to stand for a few seconds in the breeze created by one of these high-speed fans. The use of electric fans has been adopted to an extent not warranted by the results possible to attain with them, and their use will no doubt be gradually discontinued as the fallacy of the idea becomes apparent.

Primitive forced air circulation.

The first system of true forced circulation to consider is not illustrated and needs no sketch to explain the working of it, as there are practically no distributing air ducts, the cold air being forced into the room at two or three large openings, and taken out in the same way. There are two prominent houses using an air system constructed on these lines, one having the cold air inlet near floor and warm air outlet near ceiling, the other having both cold and warm air openings near ceiling. No distribution of circulating air of any consequence is provided, the idea being simply to cool the room by forcing in air which has been cooled by coming in contact with cooling pipes located outside of the rooms. The cold air is taken in at one extremity of room and the warm air out at the other, or the cold air is taken in at ends of room and warm air out at center, or the reverse. This is what may be called a primitive form of forced circulation, and is quite similar to the systems of indirect steam heating as first employed. It needs no argument to show that a room equipped in this way has varying degrees of temperature, humidity and circulation, depending on the remoteness or proximity to the direct route between cold air inlet and warm air outlet—the air moving through the area of least resistance, which is usually along the center alley of room.

Linde-British air-circulation system.

Fig. 7 shows the arrangement of ducts for air circulation used in the Linde-British air system; *a a*, cold air ducts; *b b*, warm air duct. This system of

refrigeration originated in Europe, and has found
favor to some extent on this side of the water, three
houses known to the writer being operated on this
system. Mr. E. H. Johnson describes the apparatus
used in cooling, purifying and circulating the air on
page 96 of the February, 1898, issue of *Ice and Refrigeration.* This consists essentially of a tank containing
brine, which is cooled by direct expansion piping.
Slowly revolving in the tank, with a portion of their
surface exposed above the surface of the brine, are
large metallic disks. A fan causes the air to circu-

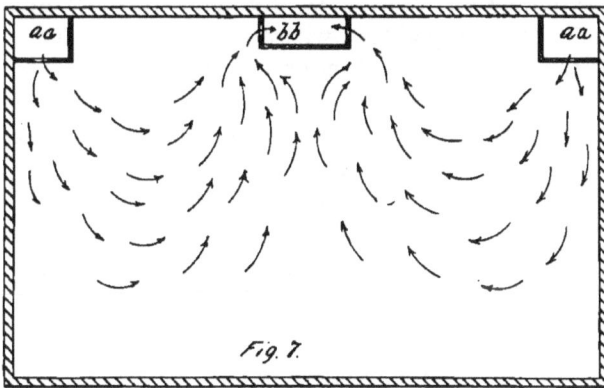

Fig. 7.

late rapidly over the brine moistened surfaces of the
disks, and the moisture is extracted, and impurities
and odors absorbed by the cold brine. In a modifica-
tion of this apparatus the direct expansion coils do the
cooling and take the place of the disks as well. The
coils are exposed to the air circulation, and the brine is
pumped over them in a shower, a shallow pan or tank
being provided under coils as a reservoir and receptacle for the brine—the same brine flowing over coils
again and again. It has been claimed for this system
that almost any product can be stored in different
rooms, all of which were fed by the same air circuit
and cooling apparatus, without any injury to the most
sensitive. A statement of this kind must first be

proven before we can accept it. There is no doubt, however, but that some good features are embodied in above described apparatus. It is well known that water, and especially salt water, has a great affinity for impurities contained in the air, and when the air is circulated in contact with the brine, as in the Linde system, many of the gases and impurities common to a storage room are absorbed. That they are absorbed to any greater extent than when the moisture is simply frozen on the pipe coils, the writer is not prepared to assert. An objection to this brine or wet surface air cooling is the liability of trouble when brine gets polluted with impurities. After some use the brine will no longer act as a purifier, and in this condition will contaminate, rather than purify, the air. If attended to, this trouble can be prevented by a periodical renewal, or by supplying a certain amount of fresh brine at regular intervals and allowing a portion to overflow. In discussing absorbents we will find a description of a device which seems to have all the advantages of the Linde method, and without some of the objections.

Air distribution with Linde-British system.

The location of air ducts as adopted by the Linde-British company seems to call for some attention. It is evident from their location that gravity is depended upon for a circulation of air near floor, as both warm and cold air ducts are placed on ceiling of room. The flow of air into room is controlled by means of sliding gates, which are adjusted to openings placed five or six feet apart in the air duct. This does not provide a well distributed circulation, as those goods directly opposite openings in cold air ducts will be exposed to a sharp blast of air, while others get practically none. As a result of placing the cold air duct on ceiling, there will be little or no movement of the air near floor, when rooms are filled with goods.

Fig. 8 shows a cross-section of a room fitted with a system of air ducts and false ceiling for the circu-

lation of air in a cold storage room, which the writer
has developed after several seasons' experiment,
and which is regarded as very nearly theoretically
perfect. In practical working it gives very superior
results, and is believed, by those who are using it, to
be in advance of any other system now in use. By
referring to the sketch it may be seen that the air is
forced to cover very uniformly the entire cross-section area of the room—a result not possible with any
other device. This is obtained by the use of a false
ceiling, *b b*, perforated at intervals with small holes, Arrangement
of air ducts in
the "Cooper
system" of
forced air
circulation.

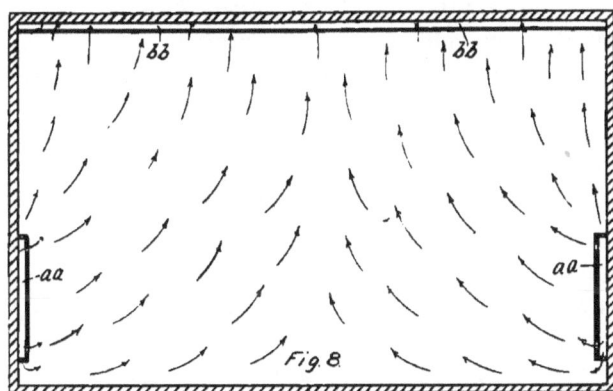

Fig. 8.

which covers the whole ceiling of room; and the side
air duct, *a a*, perforated with small holes on top, bottom and sides. The air from cooling coils is forced
into ducts, *a a*, and flows out through the perforations.
Passing through the piles of eggs, as shown by the
arrows, the air moves upward through the perforations in the false ceiling, and thence through space
between false ceiling and ceiling, to cooling coils
again. This circulation is actuated by an exhauster,
or blower, preferably located on the main cold air
duct, between the cooling coils and cold air ducts, *a a*.
This has a tendency to put the egg room itself under
a slight pressure, and the coil room under a vacuum.
In this way the air leakage from outside, if there is

any, is into the coil room, and not into the storage
room. The perforations for outflow of air from cold
air duct, *a a*, are twice as numerous on bottom as on
the top, and some are placed in the face or side of
duct also. The perforations are comparatively small
holes placed quite near together, obviating all strong
drafts, and at the same time insuring a very penetrating circulation which will not allow of any dead
corners. The false ceiling is perforated likewise, the
holes being most numerous through the center of
room at farthest point from cold air ducts, and more
widely separated as they approach sides of room
directly over cold air duct. If air ducts are correctly
proportioned, the perforations properly located and of
correct size, and eggs are piled uniformly from side
toward center of room, the air is forced to percolate
through the piles of eggs where its presence has such
a salutary effect, for reasons already discussed. The
exhauster for handling the air can be placed in almost
any location handy to power, and the air conducted to
it, but a more desirable arrangement is a direct connected or a direct belted electric motor. Then the
fan may be placed advantageously to get direct and
shorter air ducts, saving both space in the storage
room and expense in construction.

Saving of space
by using the
forced circula-
tion system.

A saving of space can be obtained by using this
system of air circulation, amounting in some cases to
fully 10 per cent of the total space cooled, and a saving of 5 per cent can be had in any house, if skill and
care are used in arranging cooling coils and air ducts.
Where rooms are moderately high the space over a
hallway or corridor may be utilized for cooling coils.
As the cooling coils are located entirely outside of the
storage room proper, the room itself can be filled
with goods snug up to the false ceiling at the top and
against air ducts on sides. The side ducts and false
ceiling only occupy about two inches of space in their
respective locations, so the space occupied by the air

system is very small. Storage men will at once appreciate that a saving of even 5 per cent in space adds that much to the profits of the season's business, as it is as expensive to cool vacant space as it is to cool it when occupied by goods.

As an objection to the forced circulation system, Cost of running fans. it has been urged that the expense of running fans continuously for handling the air was so large as to be a serious item of expense. With a well constructed apparatus and a large light weight fan wheel running at a slow speed, the air in a room containing 15,000 cubic feet, which will store about 5,000 cases of eggs, may be circulated with an expenditure of effort not exceeding one-half a horse power.

The selection of a fan for propelling the air is of considerable moment when installing an air circulating system in a cold storage room. Fans also play an important part in handling air for ventilating, and the merits of the different kinds and forms of fans will be discussed under "ventilation."

CHAPTER IV.

VENTILATION.

Value of ventilation.

IN discussing humidity and circulation, it has been explained how a large portion of the gases of decomposition and impurities of various kinds, which are incident to the presence of perishable products in cold storage, are carried by the moisture existing in the air, and that when this moisture is frozen on the cooling pipes, or absorbed by chemicals, the foul matter is largely rendered harmless. It may now be noted further that even with a good circulation and ample moisture absorbing capacity, there will still be some impurities and gases, detrimental to the welfare of the stored goods, which have little or no affinity for the water vapor in the air, and consequently accummulate in the storage room. Ventilation is necessary to rid a refrigerator room of these permanent gases. The introduction of a large volume of fresh air is not essential, however, for the purpose of purifying rooms in which eggs are stored, because the accumulation of permanent gases in an egg room is quite slow, comparatively (as in rooms where well ripened fruit is stored); but a small supply of fresh air continuously, or at regular intervals, is of much benefit.

Ventilation by the opening of doors not desirable.

This subject of ventilation for refrigerator rooms has been very much talked about recently, but about which really little is known, so far as any tangible information is concerned. Some of the more progressive cold storage managers have given some attention to this part of the business, but many of the largest and best known houses do not ventilate their rooms at all, except perhaps during the winter or spring, when rooms are aired out for the purpose of whitewashing. In some cases the change of air incident to opening and closing of doors, when goods are placed in storage or removed therefrom, is relied on to supply ventilation. This is quite inefficient, because eggs are mostly stored during two or three months in the spring, and removed from storage during the fall and winter, leaving three or four months when no fresh air of consequence can penetrate to the room, except as the doors may be opened for the purpose of taking the temperature of the room. Furthermore, this kind of ventilation during the warm weather of summer and during a large part of the spring and autumn months is worse than no ventilation at all. Some storage men even take so radical a position on this matter of opening doors during warm weather, as to insist that the door shall not be opened for the purpose of reading the thermometer. A double window is placed in the door of each room, with the thermometer hanging so that it can be read from the outside without opening the door. While the writer has not practiced this method, it seems to be a good idea, and it is certainly preferable to ventilating the room through doors which open to the outside air. When doors into rooms open into a corridor, the evil can be partly prevented by piping the corridor overhead, so that the moisture and impurities may be taken up in this way; but opening the door or window of a storage room directly to the outside air when the temperature outside is materially higher will always result in more or less bad effect on the goods, as a result of the water vapor in the warmer incoming air being condensed on same.

Another source of ventilation similar in its results Air leakage into storage room. to the opening of a door or window is that resulting from the leakage of air directly into the storage room, through the pores and crevices in the walls around the doors and windows, etc.—leakage of air literally—air that gets in when everything is supposed to be closed. The amount is usually imperceptible, but is enough in some houses to be a serious detriment to the quality of work done. In small houses with large outside exposure and poor insulation this air leakage is considerable, but in the big refrigerators of several hundred thousand cubic feet capacity, and with thorough insulation, it is reduced to practically nothing. The loss of refrigeration caused by air leakage, while of some importance, is of small moment beside the bad effects resulting from the moisture and impurities brought in by the warm air from the outside. The value of prime, tight insulation, as a conserver of refrigeration, aside from a matter of keeping out the warm, moist air, is well enough understood not to need repeating here, but a word about windows and doors is properly in line with the present discussion.

Use of windows for egg rooms.

Rather than consider what might be a good way of placing windows in a cold storage building, their use should be discouraged. Even with four or five separate glass, divided by air spaces, and with all joints set in white lead, the loss of refrigeration is large. It is also very difficult to fit insulation around the window frame so as to make a good job; and even if a passable job were practicable, the expense of putting in windows is sufficient to condemn their use. The increased fire exposure is of some consequence, too, and with the low cost of electric light, windows should not be thought of for cold storage work. Barring the small amount of heat given off, the incandescent electric lamp is an ideal device for lighting cold storage rooms, as the air is not vitiated as when using gas, kerosene or candles.

The Stevenson door for egg rooms.

Doors which will shut tight, forming a nearly perfect air seal, with a small amount of pressure, have long been wanted for cold storage rooms. Most of the ordinary bevel doors, either with or without packing on the bevel, will not shut even approximately tight; and in operation nine out of every ten stick and refuse to open except after many persuasive kicks and surges—we all know how it is. While having no interest in furthering the sale of the Stevenson door, which will be advertised in *Ice and Refrigeration*, the writer believes it to be head and shoulders above anything else in this line, and does not hesitate to recommend it to those wanting a door which will prevent air leakage. The price is very reasonable, considering the excellent material and fine work put into its construction. The slight additional cost over the common door will be quickly saved, by reason of its quick action—opening instantly when the lever is grasped.

Having got into the subject of air leakage, we may as well see how it is caused and why it must be guarded against. It is operative from the same law as gravity air circulation, which was explained quite thoroughly in the first part of the chapter on "Circulation." When the outside air is warmer than that of the storage room, the air in the storage room produces a pressure on the floor and lower part of the room, by reason of its greater weight, and consequently it seeks to escape there. If there are openings near the floor where the air can flow out, and others at the ceiling or upper part of the room, the air will flow in at the top and out at the bottom of the room. Reverse the conditions of temperature, and the direction of flow of air is also reversed. That is, when the air outside is colder than the air of the room, the cold air will flow into the room at the bottom and the comparatively warm air of the room out at the top. This action is nicely illustrated by noting the air currents in a door which is opened into a cold room when the temperature is very warm outside. The warm air rushes in at the top of door and the cold air of room out at the bottom. In cold weather the direction of air flow will be reversed. *Principle governing air leakage.*

Perfect inclosing walls for a cold storage room would be perfectly air tight, as they would be if lined with sheet metal, with soldered joints. The interior conditions would then be under more perfect control. *Walls of a perfect cold storage room.*

It is hardly necessary to do this (although it has been done in case of some old time houses), as a practically tight job may be had by using the right materials, well put on. Air leakage may not be exactly ventilation, but it is a kind of ventilation which has given the writer some trouble in the past, and does still, consequently the difficulties of operating a house with defective insulation and large outside exposure, and still turn out first-class eggs, are very thoroughly appreciated.

Air for ventilation must be dried and cooled.

Methods of ventilation which are permissible when applied to the work of supplying fresh air to ordinary structures are generally dangerous when used to ventilate cold storage rooms. The problem in ventilating non-insulated structures is merely the supplying of fresh air from the outside without causing a marked change in the temperature, and without creating strong drafts. Air for the ventilation of refrigerator rooms, during warm weather, must be of very nearly the same temperature and relative humidity as the air of the room to be ventilated, and free from the germs which hasten decay and cause a growth of fungus on the products in storage. If a door or window of a storage room is opened directly to the outside atmosphere, there will be little or no circulation of air into and out of the room when the temperature outside and in is about the same, unless the wind should be favorable. As we cannot ventilate in this way when the air outside is colder than the storage room, on account of freezing the eggs, and the introduction of fresh air, which is warmer than the storage room, is not permissible, for reasons already given, the matter reduces itself to not ventilating at all during warm weather (which most houses practice), or of properly cooling and purifying the air before forcing it into the storage room. It will bear repeating that it is positively bad practice to allow air from the outside to get into an egg room during the summer months, also during a large portion of the spring and fall months, unless cooled and purified first. The fact that we cannot see the moisture deposited in the form of beads of water, or floating in the air in the form of fog or mist, does not indicate that it is not present. The sling psychrometer, described in discussing humidity, will give an accurate indication of the result of this unscientific method of ventilating.

Any natural means of handling air for ventilation Robert Briggs on fan is inaccurate and inoperative, or it may be positively ventilation. harmful, except under favorable conditions. If depending on natural gravity for ventilation it will be guesswork, to a greater or less extent, because depending on conditions which vary with the season, temperature, direction and force of the wind, etc. The late Robert Briggs, an authority on ventilation, makes a concise statement of the advantages of using fans for ventilation, in his "Notes on Ventilating and Heating."* He says: "It will not be attempted at this time to argue fully the advantages of the method of supplying air for ventilation by impulse through mechanical means—the superiority of forced ventilation, as it is called. This mooted question will be found to have been discussed, argued and combated on all sides in numerous publications, but the conclusion of all is, that if air is wanted in any particular place, at any particular time, it must be put there, not allowed to go. Other methods will give results at certain times or seasons, or under certain conditions. One method will work perfectly with certain differences of internal and external temperature, while another method succeeds only when other differences exist. . . . No other method than that of impelling air by direct means, with a fan, is equally independent of accidental natural conditions, equally efficient for a desired result, or equally

* Proc. Am. Soc. Civil Engineers, May, 1881.

controllable to suit the demands of those who are ventilating."

There are two general methods, with some modifications, for handling air for ventilation: The plenum or pressure method, in which the fresh air is forced into the room; and the vacuum or exhaust method, in which the foul air is drawn out. The exhaust method is to be avoided for ventilating cold storage rooms, for reasons which we shall see presently. With this method, sometimes the exhaust steam from an engine is utilized to induce a draft of air upward from storage room, by heating the air in a stack or ventilation flue connected at its lower end with the room to be ventilated. In some cases no provision is made for an inflow of fresh air, in which case it will seep in at every crack, crevice and pore (by reason of the partial vacuum created by exhausting the foul air out), bringing a load of moisture and germs of disintegration into the storage room. This exhaust steam method is no different in its result than if a fan were placed so as to draw the air out of the storage room under conditions which are otherwise the same as described in connection with the exhaust steam method. Should we provide an inlet for fresh air, through proper absorbents, the same law would be operative, only to a lesser degree, as a partial vacuum must, in any case, be created before the air from outside would flow into the room, tending to the dangerous air leakage already fully discussed.

The plenum or pressure method is by far the best for our purpose. The air should be forced into the room by a fan, after first properly cooling, drying and purifying it. An outlet for the escape of the foul gases which it is desired to be rid of, should be provided near the floor, as these gases, by reason of their greater gravity, tend to accumulate in the lower part of the room. It will be observed that forcing the fresh air in creates a pressure inside the room, and if there is any air leakage, it will be outwardly from the room—exactly the way we want it to go. Having brought our subject to the point where it is found that the best way to ventilate is by the use of fans forcing the air into the storage room, we will determine what type of fan is best adapted to our needs. What is said of fans for ventilation is equally true if they are to be used for forced air circulation, described under head of circulation.

It is admitted by a majority of experts on air mov-ing machinery that the disk or propeller wheel type of fan, through which the air moves parallel to the axis of fan, is not efficient or desirable for work where the air has to travel through a series of tortuous air ducts, as in the forced air circulation system for cold storage work, or for ventilation purposes where there is some resistance. Where any resistance of importance is encountered, the disk fan must be driven at a high rate of speed, and at an immense loss of power, to compel it to deliver its full quota of air. Another disadvantage of the disk type is the difficulty of belting to the shaft, or of getting power to the fan in any form, if it is inclosed entirely in an air duct. The disk type will therefore be dismissed, and the well known centrifugal, or peripheral discharge fan taken up. *Disadvantages of the disk type of fan.*

This type of fan draws the air in at its center parallel to the shaft, and delivers it at right angles to the shaft at the periphery or rim of the fan wheel, the law governing its action being the well understood centrifugal force, which is commonly illustrated when we see the mud fly from a buggy wheel or the water off a grindstone. The advantage of these fans over the disk type is that the centrifugal action set up by the rotary motion of the fan is utilized to give velocity to the air in its passage over the fan blades. In the selection of a fan for the purpose of forced circulation in the storage room, or for forcing in fresh air for ventilation, it should be noted that a large slow *Advantages of the centrifugal type of fan discussed.*

Proportionate power consumption of fans at different speeds

running fan wheel is very much more economical of power than a small fan running at a high rate of speed, both doing the same amount of work. The loss of refrigeration, too, in a rapidly moving fan, is of consequence, because the air is warmed by impact with the blades. The proportion of power saved by the use of a large fan running at a slow rate of speed, rather than a small fan running at a high rate of speed, both delivering the same amount of air, is almost phenomenal, and does not seem at all reasonable at first view. The volume of air delivered by a fan varies very nearly as the speed, while the power required varies about as the cube of the speed. That is, doubling the speed doubles the volume of air, while the power required is increased eight times. We will take a specific case. A 45-inch fan wheel, revolving at a speed of 200 revolutions per minute, delivers, say, 5,000 cubic feet of air per minute, and requires but one-quarter of a horse power to operate it. If the speed is increased to 400 revolutions, the volume of air delivered will be only about 10,000 cubic feet, while the power required to drive it will be raised to two horse power. These figures are theoretical, but within certain limits are approximated in practice.

Loss of power from excessive fan weights.

For use in cold storage work the objection common to nearly all the air moving machinery found listed by the manufacturers is the seemingly unnecessary amount of metal used in its construction. The heavy weight of the fan wheels, and the large diameter of shaft necessitated by such weight, causes much friction on the journals, so that when running at the slow speeds desirable for cold storage work, more power is required to overcome the mechanical friction than is actually required to move the air.*

---

*Having been unable to find a fan wheel well suited to the requirements of cold storage duty, the writer has designed and constructed a line of fan wheels especially for slow speeds, which are amply strong and capable of moderately high speeds, when necessary, but are very much lighter than most fans on the market, and consume proportionately less power in mechanical friction.

No doubt the high speeds necessary for some work have obliged the manufacturers to make their fans amply strong for the highest speeds, consequently they are not economical for the slower speeds. It would not be appropriate for a person to fan himself with a dinner plate—it would do the work, but would not be economical of power.

So far we have found out what kind of ventilation is not desirable, and have an inkling of what kind would be desirable. The question before us now is to properly treat the air before introducing it into the storage room, so that it may be fresh—*i. e.*, pure oxygen and nitrogen, without excessive moisture, and free from the impurities and germs which may contaminate the product which is being refrigerated.

The questions referring to temperature contained in the letter of inquiry sent out by the writer before beginning to write these articles are as follows: Questions relating to ventilation.

First.—What plan do you pursue in ventilating egg rooms?

Second.—Under what circumstances and how often do you ventilate?

Third.—How often do you consider it advisable to make a complete change of air?

Outside of a bare dozen, the replies on this much-talked-of subject were of no value whatever for our purpose. Most of those answering do not ventilate; many others get their ventilation through the opening of doors; some ventilate through an elevator shaft, by opening doors at top and bottom, etc. Only three or four are properly cooling and drying the air before introducing it into the egg rooms. One successful storage manager says that. "It is trouble enough to take microbes, bacteria, moisture, etc., out of one batch of air" (meaning the air in his rooms at the beginning of the season), without adding to his troubles by sending in more air loaded down with the same mischief makers. As before pointed out, unless the air to be Miscellaneous replies to ventilation queries.

used for purifying the rooms is itself first cooled and purified, this man's idea is perfectly correct.

Impurities existing in the free outside air. The free outside air during warm weather, especially in the vicinity of our large cities, contains, among many others, germs which produce the parasitic plant growth which is called mildew or mold. The exhalation from the lungs of the many animals and men who inhabit our cities, and the evaporation from the dust, dirt and decaying matter of various kinds peculiar to the street, render the air a receptacle and conveyor for impurities and germs of many species. The species of germs which concern us are active in proportion to the temperature and humidity of the air. In a warm atmosphere which contains much moisture they take root and grow rapidly, throwing off more germs of their kind, which impregnate the air in an increasing ratio as the humidity and temperature are increased. The humidity of the outside air is not necessarily increased with the temperature, but it is always increased to some extent, and as the temperature of the outside air rises we must necessarily be more and more careful how we treat and handle the air which we are to use for the ventilation of refrigerator rooms.

Cooling air for ventilation. It is readily understood why it is necessary to cool the air before introducing it into the storage room to at least as low a temperature as that of the room to be ventilated, and some cold storage managers have ventilated on this basis, thinking that this was all that was necessary for successful ventilation. Air cooled only to the temperature of the storage room will be saturated with moisture at that temperature, and will be in condition to develop mold rapidly. An improvement on this manner of handling is to cool the air to be used for ventilation to a few degrees (say five or six) below the temperature of the storage room. The air will then be rendered as dry as that of the storage room. This is a good method of ventilation, and one which the writer has practiced, but it is open to criticism, because of the fact that the air is not purified fully at the same time it is cooled and dried. If the air is first cooled to several degrees below the temperature of the room to be ventilated, it will be of benefit to the room, if not overdone, but in results will not be equal to a system to be described and illustrated further on in this article.

Several houses known to the writer ventilate by Inefficient method. letting the warm outside air in at a high point of the ceiling, directly over cooling coils, expecting that the air will be properly cooled and dried before it flows into the room itself. The same objections are applicable to this system as are applicable to any plan of ventilating where the air is cooled only to the temperature of the room to be ventilated, because the air will be at the saturation point, and will therefore raise the humidity of the room, as well as introduce a quantity of germs and impurities.

If we ventilate by simply cooling the air, the Simple air cooler. simplest and most effective method, as shown in Fig. 9, is to take the air from as high and sheltered a place as is accessible about the building; draw it down over frozen surfaces in the form of brine or ammonia pipes, which may be arranged anywhere along the wall of a room, outside of the storage entirely, if more convenient. An exhaust fan takes the air from the coils in the ventilating flue and forces it into the room to be ventilated, allowing it to escape in the neighborhood of the cooling coils, where it will mix with the air circulation, and flow into the room through the regular channel. It is necessary to provide an outlet for the escape of foul air whenever fresh air is forced into the room. This outlet should be near the floor, and of about the same area as the inlet pipe. A steam coil may be provided beneath the cooling coil in ventilating flue, as shown in the sketch, for the purpose of melting the frost off the

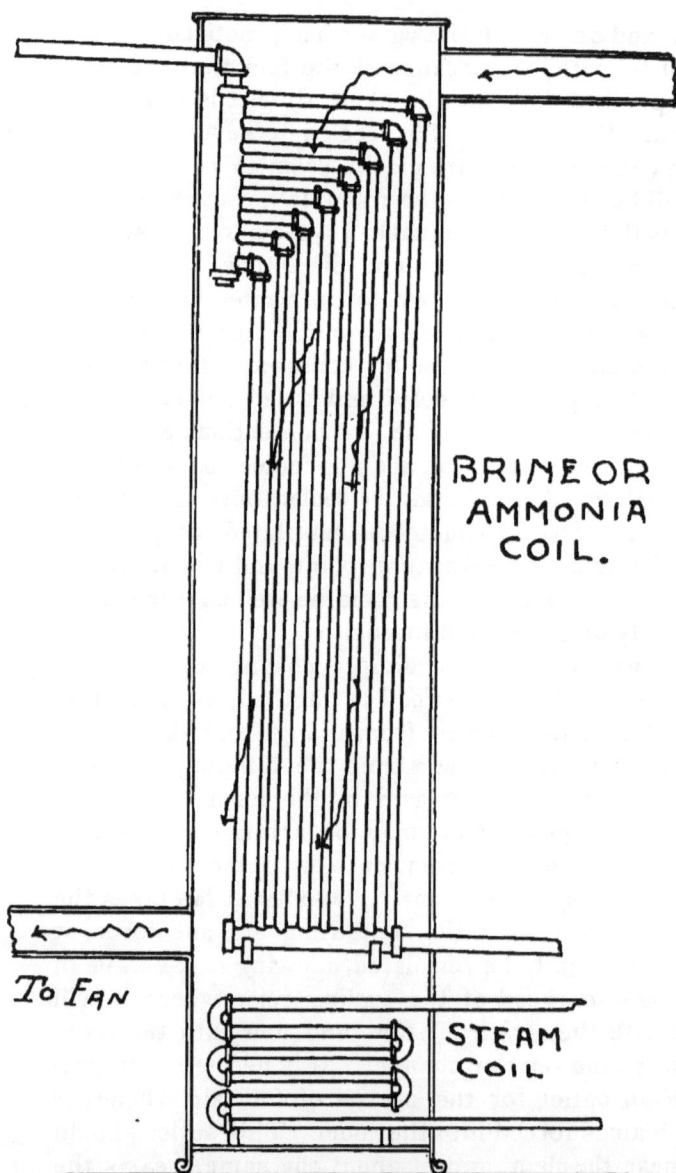

BRINE OR AMMONIA COIL.

TO FAN

STEAM COIL

FIG. 9.

pipes. The casing around the cooling coil should, of course, be insulated moderately, as well as the pipe leading from it to the storage room, wherever exposed to the warm outside air. The size of apparatus necessary for this purpose need not be large, as the quantity of air necessary for ventilating egg rooms is quite small, comparatively.

"Americus" mentions a method of washing air for ventilation, in the July, 1898, number of *Ice and Refrigeration*, which seems to have advantages. The idea is to draw or force air through a body of water or brine by immersing the intake pipe so that the air will bubble up through the liquid. This seems quite simple, but when it comes to forcing air through a liquid with a fan it is not so simple, as nothing short of an air pump will drive air through a pipe submerged as above described, unless the opening from pipe is placed quite near the surface of the liquid; in which case the benefit to the air is very small. Experiments conducted by the writer along this line were considered failures.

Shown in Fig. 10 is what appears as a rather complicated apparatus, but on investigation it proves to be quite simple. There are three members to this system, as follows: *First*, The air washing tank, in which the air flows upward against a rain of water from a perforated diaphragm above, as clearly shown in the sketch. This not only cools the air to the temperature of the water, say 55° F. or 60° F., but it also takes out a large portion of the impurities of various kinds. From the washing tank the air is passed on, in a comparatively pure and cool state to be still further cooled. *Second*, The cooling tank, in which the air is cooled to several degrees lower temperature than that of the storage room. This removes the moisture which holds in suspension the few impurities which may have passed the washing tank, the moisture being deposited on the frozen surfaces within the cooler. From the cooler the air is passed into, *third*, the drying box, which contains chloride of

[margin note: Submerged intake method.]

[margin note: "Cooper system" warm weather ventilation.]

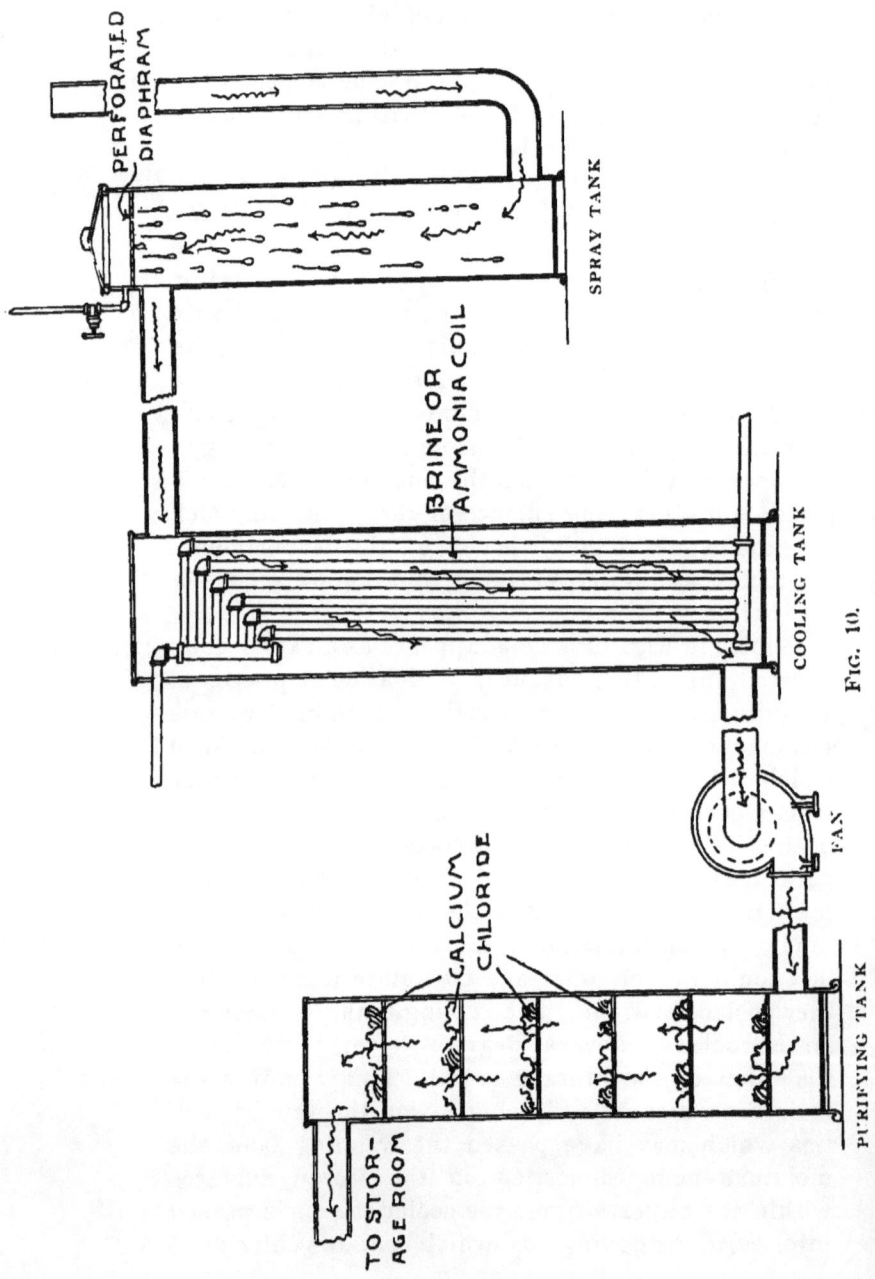

PERFORATED DIAPHRAM

SPRAY TANK

BRINE OR AMMONIA COIL

COOLING TANK

FAN

CALCIUM CHLORIDE

PURIFYING TANK

TO STOR-AGE ROOM

FIG. 10.

calcium. This chemical is a well known absorber
of moisture, what is technically known as a deliquescent substance. If moisture of any account passes
the cooler it is surely stopped in the dryer, which
"makes assurance doubly sure," so far as delivering
a pure, dry air is concerned. The " microbes, bacteria, moisture, etc." (which influenced the gentleman mentioned previously not to ventilate), are effectually disposed of by this method. It would be a
hardy germ, indeed, that would not succumb to such
vigorous treatment.

The volume of air necessary for ventilating a Volume of air
given size of egg room can only be estimated, and required.
probably no two storage men will agree as to what is
a correct quantity. Some say that the introduction
of a volume of air equal to that of the room to be
ventilated should take place each day; others twice
each day; some even take so radical a view of it as to
say the oftener the better if the air is properly dried
and cooled. This is of course true enough, but the
foul gases which we can be rid of by ventilation accumulate but slowly in an egg room, and it is probable
that the introduction of a volume of fresh air, properly
treated, equaling that of the egg room, twice each
week will be ample for the purpose of keeping the
room in good condition, and in most cases once each
week may do nearly as well. There is much to be
developed yet in the direction of ventilation of refrigerator rooms, more particularly in the way of some
method of knowing when a room requires ventilating.
Perhaps Prof. Siebel or some equally bright chemist
may be able to assist us on this point by informing us
what the gases are which we must dispose of, and
indicate some simple method of determining their
presence, and in what proportion.

All that has been said about ventilation so far
applies only to the ventilation of cold storage rooms
when the air without is warmer than the air of the

storage room. We will now give our attention to another kind of ventilation, that is applicable when the air without is at about the same temperature as the storage room, or at some degree lower. This will be designated as cold weather ventilation, as this term seems to express its function perfectly.

Natural ice cold storage in cool weather.

It has long been a well understood fact that eggs and other products held at about the same or a higher temperature take more harm in cold storage during the cool or cold weather of fall and winter than during a long carry through the heated term. Much has been said and written about why the old style overhead ice cold storages give such poor results during fall and winter, the reason assigned being lack of circulation, as the meltage of ice ceases when the cool weather comes. This is true; further, the large body of ice becomes an evaporating surface, and the dirt and impurities which are found in all natural ice, to a greater or less extent, have accumulated on the top of this ice, and the evaporation which takes place carries gases from this miscellaneous matter into the air of the storage room, with consequent bad results. In some houses this may be avoided by closing the trap doors covering circulation flues, but it is seldom done, and in many houses it is impossible.

Pipe cooled rooms in cold weather.

Now are we who cool our storage rooms with brine or ammonia pipes very much better off in this one respect than those who have these much despised overhead ice cold storages? Our rooms are cooled by frozen surfaces, on which accumulates the evaporation from the goods in store, which, as we have already plainly seen, contains much foul matter and impurities. Precisely as in the ice cold storages, the cooling surfaces, which absorb moisture during warm weather, become evaporating surfaces, and give back to the air of the room a considerable portion of the various impurities and germs which have been accumulated during the warm weather of summer. To make this point

more plain it may be considered thus: During the
period when the outside air is considerably warmer
than the air of the storage room it is necessary to keep
some refrigerant at work cooling the air within. This
is usually done by circulating brine or ammonia
through pipes, and the air of the room is circulated in
contact with the pipes. When the outside temperature is high, more of the refrigerant must be
circulated, or its temperature must be lowered;
as the weather turns cooler in the fall, less refrigerant, or the same amount at a higher temperature, must be circulated, and when the air without reaches the temperature of the room, the
circulation of refrigerant must be discontinued altogether. When this is done the moisture on the
cooling pipes begins to evaporate. This evaporation
added to that which is given off by the eggs themselves soon renders the air saturated with very impure and poisonous vapors, which cause the eggs to
deteriorate very rapidly.

The influence which the temperature of the refrig- Influence of
temperature of
erant flowing in the cooling pipes has on the condition refrigerant in
cooling pipes.
of a storage room may be better understood by taking a specific case: A room with a temperature of
33° F. and a humidity of 70 per cent has a dew point
(temperature at which the air precipitates moisture)
of 25° F. Therefore any cold surface (as a pipe surface), having a temperature of 25° F. or lower, will
attract moisture when exposed to the air of the room.
If the pipe surfaces are heavily coated with frost, as
they usually are as cold weather approaches, the frost
acts as an insulator, and the refrigerant flowing in
pipes must be at a considerably lower temperature
than the air of the room, or no moisture is attracted.
We have all noted how the accumulation of moisture
on pipe coils is slower and slower as the thickness increases, until finally a limit is reached where no more
frost will form; yet owing to the largely increased

surface the room can be kept at its normal temperature. If pipes are badly loaded with frost, sometimes
no absorption of moisture will take place when the
refrigerant flowing in the coils is 10° or 15° below the
temperature of the room. The surface exposed to
the air of the room, whether in the form of frost or
otherwise, must be at or below the temperature of the
dew point, or no moisture will be absorbed. The
value of suitable moisture absorbing surfaces as the
cool weather of fall and winter approaches cannot be
overestimated, as many have found to their sorrow
that two weeks' stay in cold storage under bad conditions in cold weather will do more harm to the eggs
than four months during hot weather.

The remedy for this trouble is found in keeping
the air of the room from coming in contact with the
poisonous frost which has been accumulated on the
pipes during their period of duty during warm
weather; or what is still a better way is to not allow
the frost to accumulate on the pipes at all, by using a
device, described further on under head of absorbents. How to keep the air from contact with the frost
on pipes is not an easy matter, and in case of piping
suspended directly in the room it is an impossibility.

Evaporation
from frost
accumulated on
cooling pipes.
With a system of screens arranged around coils,
as described in the first part of the paper on circulation, trap doors may be fitted to the openings and the
air circulation shut off in this way; but the simplest
and best way is to equip the rooms with forced circulation, and locate the pipes outside of the room entirely. Then it is only a matter of shutting off the
circulation over coils, allowing it to continue through
a by-pass, or if the device shortly to be described is
used, the circulation may be allowed to continue over
coils. It seems quite clear, from what has been written, why a storage room gets foul quickly during cool
weather, and also that the bad conditions may be bet-

tered by cold weather ventilation. The harm resulting from the foul evaporation from frost on cooling
pipes may be obviated by not allowing contact between
it and the air of room, but the evaporation from the
eggs themselves must be taken up by other means
when cooling surfaces are no longer operative.

By carefully observing conditions a storage room Handling of
may nearly always be kept in prime condition during ventilation.
cold weather by no other means than the introduction
of fresh outside air at as frequent intervals as right
conditions of temperature and humidity will permit.
It is quite safe to force in plenty of air which has
about the same temperature and humidity as the
room to be ventilated. There are few impurities in
the clear, crisp air of a bright fall day, and many such
are available for our purpose in the latitude of Min-
nesota and New York, and a somewhat smaller number, perhaps, in the latitude of Iowa or Ohio. It is only
a matter of handling the free air of heaven under-
standingly. One's impressions, however, will hardly
do in judging what air is good to use for ventilating
purposes. If you have a bright, clear day, or, what is
still better, a clear, cold night, which has the appearance of being what you want, get out your sling
psychrometer and set all guesswork aside. It is
frequently possible to fill your egg rooms with fine,
pure air at a temperature about the same as that of
the room, as early as the latter part of October, if
you are watching for the opportunity. Provide a good
big fan wheel, which will handle a large volume of air
in a short time, and when conditions are right blow
your rooms full of it. Repeat this whenever the
weather conditions will permit.

We may now consider cold weather ventilation Method for
under another condition, viz.: When it is colder out- weather.
side than inside the storage room. Whenever the outside air is 8° or 10° below that of the storage room it
is always perfectly safe to introduce it into the stor-

age room, after it has been first warmed to the temperature of the room to be ventilated.   That is, it is safe so far as introducing moisture or impurities is concerned.   If we should ventilate in this way continuously our humidity would be lowered to a point where the eggs might suffer from evaporation.   It is necessary, therefore, that observation of the humidity of the room so ventilated be taken, so that this kind of ventilation may not be overdone.

Manner of introducing air.   The method of getting air into the rooms under these last two systems of ventilation is of no special moment, except that it be under control, and we have already noted that the only good way of handling air was by the use of fans, preferably large and of light weight, and running at a slow speed.   Where the forced circulation is installed, it is sometimes practicable to so connect the fans used for this purpose, that cold weather ventilation may be handled by them; but a separate fan is much better, and while seeming more complicated is really simpler to operate, because handled independently.   When using an independent fan or when using the forced circulation fan for ventilating, the fresh air mixes with the circulation and is well distributed by it to various parts of the room.

The ventilation of cold storage rooms is not a matter which can be safely left to such help as may be at hand, and if good results are to be secured "the boss" should see to it himself.   Cold weather ventilation, especially, must be handled scientifically or trouble may result instead of benefit.   No absolute rules can be given for handling ventilation because of widely varying conditions, but if what has been written is read and studied carefully the subject can be taken up intelligently and followed out to its legitimate conclusion.

CHAPTER V.

ABSORBENTS.

THE use of absorbents in cold storage rooms has Purifying air by absorbents. been common since the industry was in its infancy; their use originating, no doubt, from an appreciation of the fact that the air of a storage room quickly became too moist and impure to do the work of preservation perfectly. When absorbents and ventilation are applied to refrigerator rooms they practically have one duty in common—that of purifying the air. Ventilation purifies by furnishing pure air which displaces the foul air; absorbents by attracting the moisture, and with it the impurities of the storage room; but where ventilation is largely for the purpose of forcing out the permanent gases or impurities which have little affinity for moisture, absorbents are for the purpose of taking up the moisture and the germs and impurities which are absorbed by it.

Active absorbents can be made to perform duty in Prof. Nice's system which utilizes an absorbent. absorbing the moisture which is usually condensed on the cooling coils, as illustrated in one style of the antiquated overhead ice cold storages. If the writer remembers correctly, the system is called Prof. Nice's system. In this system the ice is supported above a water tight sheet iron floor which forms the ceiling of the storage room, the air of the room being cooled merely by contact with this cold metal surface, which is cooled by the ice above. The moisture given off by the eggs in storage, and that resulting from air leakage was taken up by an absorbent, chloride of calcium being the chemical mostly in use for this purpose. It was applied by suspending it in pans at the ceiling of the room, or in some cases on the floor under the goods. Prof. Nice's system gave good results years ago in competition with the Jackson, Dexter, McCrea, Stevens, etc., systems of overhead ice cold storage, which low

71

temperatures, and the improved systems of air circulation now in use have rendered obsolete to a greater or less extent. Mention is made of this system, not as recommending it, but to show the possibilities of absorbents in drying and purifying egg rooms.

Queries relating to absorbents.

The letter of inquiry sent out by the writer contained three questions referring to absorbents, written with an idea of ascertaining the coating used for the walls of a storage to the greatest extent; what absorbent was the favorite, and in what manner applied. The questions are as follows:

I. Do you use an absorbent or purifier in your egg rooms?

II. In what way do you use or apply them?

III. Do you paint or whitewash? What kind and how often applied?

Whitewash a good wall coating.

The most common wall coating in use for egg rooms is plain every-day whitewash, in various proportions of lime and salt. Several recommend one part of lime and one of salt. This makes a very good whitewash, giving a firm, hard surface, but unless some method of blowing warm, dry air through the rooms is feasible, it will dry very slowly, which is likely to cause it to have a mottled appearance instead of the pure white which gives a storage room such an attractive appearance. A better proportion for ordinary cold storage work is two parts of lime and one of salt. This mixture will dry faster, and will give a white surface which will not easily rub or flake off. There are many formulas for good whitewash, some of them so complicated as to be impracticable; but plain lime and salt, with perhaps the addition of a little Portland cement, will be good enough for our purpose.

This last formula would then be six parts white lime, three parts salt, one part Portland cement. In preparing this wash, proceed as follows: Slake the quicklime by pouring on boiling water, stirring

thoroughly during the process. A half bushel of The best formula for whitewash.
lime is all that can be handled easily. Pour on only
a little water at first, adding more as the mixture
thickens, and do not allow the lime to become dry, or
it will "burn" and become lumpy. When the lime is
thoroughly slaked and reduced to the consistency of
cream, add the salt while the mixture is still hot—the
salt will dissolve better—adding more water as is
necessary to keep it to the proper consistency. The
Portland cement should be added only to each pailful as used, as it sets if allowed to stand, and does
not retain its tenacity. A good sized handful to each
pailful of the wash is about right. By the addition of
a teaspoonful of ultramarine blue to each pailful, the
brown effect resulting from the addition of the cement
will be neutralized.

Storage rooms should be whitewashed during Whitewashing storage rooms.
cool, dry weather, with the doors open, or warm, dry
air from a steam coil should be circulated through
the room. This is quite a simple matter where a
house is equipped with forced circulation. Cover the
walls, ceiling and floor with a coat of whitewash
each spring, and allow ample time for the rooms to
air and dry out before goods are placed in them. If
whitewash is to have a nice white appearance it must
not be too long in drying; on the other hand, if dried
too quickly it will flake or cleave off more readily.
The quickest method of applying whitewash is with
a compressed air spray. It will make a fair job, and
is done much quicker than by hand. .

The cold water paints, which are now quite com- Cold water paint.
mon under various names, are good for many places
where whitewash will not do, as on doors and in the
corridors, or wherever the clothing may come in contact with the woodwork, or where a product is handled
which may be injured by the flaking off of whitewash.
Whitewash will generally rub or flake off to some
extent, but the best of these cold water paints are

nearly as impervious as so much oil paint, and quite valuable for nearly all interior and some exterior work. Many houses use nothing else for their refrigerator rooms, but the expense is not warranted, as whitewash will do equally well in most places. It is a good idea to keep some of this cold water paint on hand, and apply it at intervals to doors, etc., when they become soiled by handling. This is much better than to paint doors some dark color so they will not show soil—nothing compares with a pure white—and oil paint has no place about a storage room.

Shellac for wall coating.

Shellac is an old stand-by finish for refrigerator rooms, and if selected ceiling is used, it makes a very neat piece of work, as it brings out the natural grain of the wood, than which there is no more beautiful finish. The surface scratches easily and will look mussy unless renewed frequently, but there is no serious objection to shellac (barring the expense), as it is strictly odorless and waterproof. It has no purifying or disinfectant properties like lime and salt whitewash, in appearance is very little superior, and the much increased expense makes it very little used at present. Many other preparations are in use under various names, but whitewash is as good as any of them, with the exception previously noted.

Lime as an absorbent of moisture and impurities.

The two chemical absorbents in general use for taking up moisture and the impurities from cold storage rooms are chloride of calcium and lime (either unslaked or air slaked, or in the form of whitewash, as before mentioned.) Occasionally waste bittern from salt works is used, but the active principle of bittern is chloride of calcium. Ordinary quicklime has the property of absorbing moisture and impure gases from the air, and is used in very much the same way as chloride of calcium; that is, it is placed around the room on trays or pans. Lime, however, has very little capacity for moisture as compared with chloride of calcium, and when exposed to

the air it will simply air slake, which means that it
will absorb moisture enough from the air to disintegrate into the form of a powder. Lime in this form is
known as air slaked lime, and is used to a large extent in egg rooms. Air slaked lime as it comes from
the lime house will absorb very little moisture, but it
gives off minute particles of lime which have a good
effect in preventing the growth of fungus, which we
have already fully discussed. Air slaked lime is
usually applied by spreading on the floor of the room,
between the 2×4's (which are used at the bottom of
each pile of eggs), to the depth of an inch or more.
This must necessarily be done when the eggs are
piled, and consequently its efficiency is very low
when the cool weather of fall comes. This defect
has. been overcome by scattering fresh air slaked
lime through the rooms so as to create a cloud of lime
dust, but this is objected to because it musses up the
cases. A better way of using lime is in the lump
form—quicklime—which can be placed around the top
of the room in trays or pans and renewed from time
to time through the season.

Chloride of calcium is the most vigorous absorbent (or drier, as it is called) which we are discussing.
It is the same salt of the metal calcium as common
salt (chloride of sodium) is of the metal sodium. Both
have a strong affinity for water, but chloride of calcium is much the more energetic of the two. Where,
in a moist air, common salt simply attracts enough
moisture to become damp, chloride of calcium will
absorb enough water to lose its solid form entirely,
uniting with the moisture of the air to form a solution
or brine. The strong affinity of this salt for water
has been utilized for the purpose of drying and purifying refrigerator rooms, and in this capacity has
been a general favorite for years. The most primitive method of applying it is to place it in a simple
iron pan, allowing the brine to run off into a pail as

Chloride of
calcium as an
absorbent.

fast as formed. A better way is to support the calcium on a screen of galvanized wire, with a galvanized pan below for catching the brine. This allows of a free circulation of air around the calcium. This apparatus should be suspended near the ceiling of the room, one end slightly higher, to allow the brine to run off into a galvanized iron pail, supported at the low end of the pan. Galvanized iron is specified because black iron rusts badly when exposed to the air.

The writer's method of applying chloride of calcium. A still better way of applying chloride of calcium, which makes the calcium do two separate and distinct duties, is a method recently designed by the writer. It has proven itself especially valuable for use in rooms where the air is circulated by mechanical means, and in which the air is cooled by passing over brine pipes, through which the brine is circulating at a comparatively high temperature. If low brine temperatures are used in combination with the device about to be described, the tendency may be to dry the air to such an extent as to be detrimental, and ample pipe surface must be provided and the quantity of brine flowing through pipes reduced so that no excessive drying effect will take place. The device consists simply of supporting a quantity of chloride of calcium above the cooling coils, so that the brine, resulting from the absorption of moisture by the calcium, will trickle down over the pipes. This effectually prevents any formation of frost on the pipes, and therefore keeps them at their maximum efficiency at all times. The air, in passing over the brine moistened surface of the coils is purified, and as the brine, after falling to the floor of the coil room, goes to the sewer, no contamination can take place.

Do not in any method of using chloride of calcium evaporate the water from the brine and use the salt over again. The impurities will stay in the salt to a large extent, which is quite harmful, and the calcium has at least lost its value as a purifier, to a large extent. The quantity of calcium necessary depends on the conditions under which it is to be used, but in any case it is safe to use much more than the writer saw in use in an eastern house recently. A room about 30×50 and about fourteen feet high had the refrigerant shut off, and the room was in rather bad condition as to moisture, etc. In each end of the room a pail was placed, on which rested a wire screen, with perhaps ten or fifteen pounds of chloride of calcium on it. Electric fans were playing on the calcium, which was doing its best, but it seemed "like trying to dip the sea dry with a clam shell." This room should have had at least two drums (about 1,200 pounds) at work in it to do it justice.

Quantity of chloride of calcium necessary.

PACKAGE.

EGGS are continually giving off moisture from the time they are first dropped by the hen until they disintegrate, unless sealed from contact with the air, and we can therefore never hope to keep them in cold storage for several months without their losing some weight by evaporation. To prove that eggs must evaporate, the following experiment was tried by the writer in his early experience: An ordinary 30-dozen egg case was lined with tin, with all joints carefully soldered. The eggs were then placed in the fillers in the tin lined case in the usual way, and an air tight tin cover soldered on, forming a hermetically sealed package. After about sixty days' stay in an ordinary refrigerator the tins were unsoldered. The result noted was peculiar and startling. The inside of the tins was dripping wet, and very foul smelling, and the eggs were all rotten. This same experiment was tried by a friend, working independently and without knowledge of the writer's experiment. He used an ordinary fruit jar, with screw top fitting onto a rubber ring. His results were similar. In addition this gentleman packed some eggs in flour in a fruit jar, otherwise under the same conditions as the other experiment. The eggs packed in this way were all found to be in good condition when the jar was opened, as the moist evaporation from the eggs had been taken up by the flour. These experiments prove beyond a doubt that an egg must evaporate continually, and they prove further that the eggs must be surrounded by some medium which will absorb this evaporation.

We have noted in the chapter on "Circulation," how the air is best circulated so as to remove the

78

moisture and impure gases from the vicinity of the
eggs. This must be done, otherwise the fillers and
package containing the eggs would shortly be in as
bad condition as the fillers in the experiment just
mentioned. The theory and explanation of the other
conditions in the storage room necessary for successful egg refrigeration have also been taken up under
the various heads. We will now look into the requirements of the package containing the eggs while in
cold storage.

The questions contained in the letter of inquiry *Queries relating to package.*
relating to the egg package are as follows:

1. What egg package have you found to turn out
the sweetest eggs?

2. Have you used any kind of ventilated egg case,
and with what results?

3. Have you ever used open trays or racks, and
with what results?

As many different people have experimented with *Various woods in use for cases.*
different packages, hoping to get something which
would turn out perfectly sweet eggs, with little evaporation, the replies received to the questions relating
to packages are interesting, and many contained information valuable as data. The favorite package is
the ordinary 30-dozen egg case, made of whitewood, using the so called odorless fillers. The term
whitewood is usually meant to include either poplar,
cottonwood or basswood, but two or three other vari-
eties of wood, not so well known, are designated as
whitewood. Basswood is by some not placed in the
whitewood list, but the best authority known to the
writer says that basswood is as properly a whitewood
as poplar or southern whitewood. Poplar and cottonwood are most in use for storage purposes, and many
insist that basswood is objectionable because of its
liability to ferment or sour and cause tainted or musty
eggs. All kinds of cases have been in storage in the
house operated by the writer, and if all were thor-

oughly dry, no difference could be noted in the carrying qualities of the different kinds of whitewood, and the preference has been for well seasoned basswood cases. It may be that basswood is more likely to sour and affect the eggs than poplar or cottonwood, but it is always advisable to get stock for egg cases in the fall and have them nailed up during the winter, allowing two or three months for the cases to season before the opening of the egg storing term. Some have dry kilns for cases, but a naturally seasoned case is to be preferred, as then it has a chance to deodorize as well as dry out. In some localities other woods are used for egg cases. Ash, maple, hemlock and spruce have been used for storage cases, generally because they are cheaper than whitewood in that locality. Any strong scented wood like pine will not do because of the flavor imparted to the eggs.

Various kinds of fillers in use. The pasteboard frames and the horizontal dividing or separating boards which form for each egg an individual cell in the case are usually spoken of as fillers. For years only one grade of these was made —those of ordinary strawboard. When moistened by the evaporation from the eggs this material has a peculiar rank odor, which was taken up to some extent by the eggs if they were allowed to remain in the fillers for several months. Much of the flavor resulting from a growth of fungus has been laid to the fillers, and much of the flavor resulting from fillers has been laid to a growth of fungus or must, but there is no question but what strawboard fillers are not the thing for cold storage use. Many kinds of fillers have been tried, and many ideas suggested for the improvement of cold storage eggs. A white wood pulp filler made its appearance some years ago, but did not come into general use. After being in storage a few months, it absorbed moisture to such an extent as to be very soft, and they were objectionable on this account. A good manila odorless is now on the market which is giving good satisfaction where tried. Ordinary strawboard fillers have been coated with various preparations, shellac, paraffine, whitewash, etc. Any substance in the nature of waterproofing might better be left off for the reason, as we have seen, that eggs must evaporate, and a waterproof filler would hold the moisture and not allow it to escape into the air of the room. It is essential to the well being of an egg that it should evaporate, as proven by the experiments in hermetically sealing, before described. Many have gone to the expense of transferring the eggs into dry fillers in the middle of the season. One season of this was enough for the writer. A better way is to decrease the humidity of the room as the fillers become more and more loaded with moisture. The humidity may be decreased by the use of absorbents or by ventilation, as already discussed in their proper places. Fillers made of thin wood have been used in years gone by with fair success, but their manufacture has now been entirely discontinued. They were made of maple, shaved very thin, and were a prime filler so far as odor was concerned, but in cold storage the frames warp badly, and the time and eggs wasted in getting the eggs out of the fillers was a serious item against their use. As a shipping filler they were also a failure because of the excessive breakage. Some years ago an eastern company began the manufacture of what is known as the odorless fillers. These fillers are light brown or buff in color, and from the best information the writer can obtain, are composed largely of scrap paper stock, with some long fiber like manila added for strength. In the manufacture the Odorless fillers. stock is treated to a thorough washing and deodorizing process, and the result is a filler with very little odor. Eggs put up in these so called odorless fillers and subjected to the same conditions as a similar grade of eggs packed in common strawboard fillers, generally come out of cold storage markedly superior.

A number of imitations of the original odorless filler
are now on the market, some of them almost if not
quite the equal of the original. Another filler which
has given good results is the fiber filler, which is made
from a material similar to the now well known fiber
ware. They have very little odor, and remain hard
and firm while in cold storage. A new odorless filler
made from pure spruce pulp has been put on the
market this season. This is a beautiful appearing
filler, and unless appearances and the ordinary tests
are deceptive will make its mark after a trial of a year
in cold storage to prove what it can do. A ventilated
filler made by a well known creamery supply house,
has been suggested as an ideal filler for cold storage,
but they are so poor mechanically that they are not
to be thought of. The material cut away to form
the air circulation space weakens the structure of the
filler to such an extent as to make it dangerous as a
shipping filler. Whatever filler is used, it should fit
the cases, not crowding in, nor still so loose as to
shake. If this point is looked after much breakage
and consequent poor results from storage in the cold
room may be avoided.

Ventilated
vs.
tight cases.

Many styles of ventilated egg cases have been
placed on the market in years past, but very few or
none survive the test of time. A ventilated case, made
by having the sides cut an inch narrower than the
ends, has come into use, especially in one large eastern
city. Making the sides narrower forms a space of
half an inch on both sides of case at top and bottom,
for the ready access of air to the interior of the case.
This case is of very simple construction, and efficient
in allowing a free circulation of air into the case.
Others, however, prefer a case with sides in two
pieces, claiming that the cracks will allow enough air
circulation. Still others prefer the shaved or veneered
cases with solid sides and bottom, claiming that this
kind of a case will prevent excessive evaporation

from the eggs. As pointed out elsewhere in these
articles, humidity and circulation have much to do
with the evaporation from eggs; in fact, are more of
a ruling factor than the package, although the package
necessarily has much to do with it. A tight package
will allow of less evaporation than an open one. In a
very dry room with a vigorous circulation a moderately tight package is the thing, but in a comparatively moist room with poor circulation the more open
the package the better.

An appreciation of the poor circulation and damp *Storing eggs in trays or racks.*
air of the overhead ice systems has caused many of
their operators to resort to the use of open trays or
racks for the storage of eggs. Very palatable eggs
have been turned out in this way, but the use of trays
in any ammonia or brine cooled room would lead to
very excessive shrinkage of the eggs and consequent
heavy loss in candling. On a commercial scale, too,
the storing of eggs in trays is hardly practicable, as
it increases the risk of breakage immensely, and the
eggs must be transferred from the cases when received at the storage house, and back into cases again
when shipped, involving much labor, and perhaps loss
of valuable time at some stages of the market. In any
but a very moist room, eggs stored in open trays, in
bulk, will lose much from evaporation, and the loss
will be proportionately higher than on an equal grade
of eggs stored in ordinary cases and fillers. The advantage of trays, if any, for some houses, is that contamination from fillers is avoided, and about 40 per
cent more eggs can be stored in a given space. The
eggs are, however, more liable to must as a result of
moisture condensing on their surface with change of
temperature, or on the introduction of warm goods
into the storage room.

The material used for forming a cushion in the
case on top and bottom of the fillers to protect the
eggs from contact with the case, and so that they will

carry in shipping, is generally either excelsior, which is finely shaved wood, usually basswood, or the chips made in the manufacture of corks, known as cork shavings. The big cold storages recommend cork now in preference to the best excelsior. Here again comes a question of dryness. If the excelsior has been in stock for a year and stored in a dry place it is to be preferred to cork shavings, otherwise cork is the best, because we know cork is always dry. Cork makes a very poor cushion as compared to excelsior; it is liable to shift in the case, leaving one side without protection. As a matter of cost too, cork is much more expensive than excelsior. A company known to the writer manufacture a beautiful grade of basswood excelsior, which is always fairly dry when received, and makes as fine a cushion for protecting the eggs as can be desired. If people want cork in their cases they can have it by paying the price, but dry, seasoned, fine basswood excelsior is better, for reasons stated.

Eggs have been packed in oats for years, but the practice has gradually fallen off, as eggs stored in cases from the best cold storage houses have been improved in quality from year to year. Oats, if dry, will absorb moisture from the egg quite rapidly, and are objectionable on this account. If the oats are not dry the germs of mold are developed rapidly, and as the moisture is given off by the eggs, the mold will grow, causing the eggs to become "musty." Therefore the main difficulty in using oats as packing for eggs in cold storage is to have them at the correct degree of dryness. It is almost impossible to have them in the same condition at all times. Oats have also been used in cases inside the fillers, that is, the layers of eggs are first put into the filler; then the oats are sifted into the spaces around the eggs flush with the top of the filler. This is repeated through the whole case; all the space in the case not occupied by the eggs being filled with oats, excepting the small space taken by the fillers themselves, the object being, of course, to prevent the "fillers taste."

At intervals we read of some method of preserving eggs, which is said to be sure to supersede ordinary cold storage for the good keeping of eggs. A scheme was tried on a large scale somewhere across the water, in which the eggs were suspended in racks in a cold room—the racks being turned at regular intervals by automatic machinery to keep the eggs from spoiling, that is, to keep the yolk from attaching to the shell. A low temperature will prevent this, as pointed out in the chapter on temperature, and why a man should waste good energy inventing such a machine is passing all comprehension. The quantity of various chemical preparations manufactured and sold for egg pickling or preserving is even now quite large, but the high class stock now turned out by the best equipped cold storage houses has made any other method of preserving eggs at the present day almost entirely obsolete.

Complicated and obsolete methods of preserving eggs.

CHAPTER VII.

REMARKS.

Some "dont's" on egg packing, handling and storing.

THERE is a long string of "don'ts" in regard to packing, handling and storing eggs which might be put down, but the writer will be content with a few of the simpler and most useful ones. To start with, don't store very dirty, stained, cracked, small or bad appearing eggs of any description. Have your grade as uniform as possible. The culled eggs will usually bring within two cents of the market price, and it pays better to let them go at a loss rather than try to store them. Don't use fillers and cases the second time; they are more likely to cause musty eggs than new ones. Don't ship eggs in cold cars, or set eggs which are intended for storage in ice boxes. In shipping eggs from the producing section to the storage house in refrigerator cars, no ice should be put in the bunkers, because if the eggs are cooled down and arrive at their destination during warm or humid weather they will collect moisture or "sweat," and an incipient growth of mold will result. Don't use heavy strawboard fillers for storing eggs. If "the best way to improve on a good thing is to have more of it," then the best way to improve on a poor thing is to have less of it; and if strawboard fillers are objectionable, then the thinner they are the better, because less of the material is present to flavor the eggs. Further, the thin board fillers are more porous, and allow of a freer circulation of air around the egg. As already stated, odorless fillers are better than any strawboard fillers. Don't use freshly cut excelsior. It should be stored in a dry place at least six months. Use no other kind but basswood or whitewood. Don't store your cases, fillers or excelsior in a basement or any damp place. Don't run warm goods into a room containing goods

already cooled when it can be avoided. For this reason very large rooms are not to be desired. A small
room may be quickly filled with goods and closed until
goods begin to go out in the fall. If a large room is
used it may require several weeks to fill completely,
during which time the fluctuation of temperature is
at times excessive, causing condensation on the goods,
which will propagate must quickly.

To illustrate: We will suppose the egg room partly Don't put
filled with goods cooled to a temperature of 30° F. Sev- a cold room
eral cars of eggs at a temperature of, say, 70° F. are run already cooled.
into the same room. The new arrivals, in cooling to
the low temperature, give off large quantities of vapor
from cases, fillers and the eggs themselves, the vapor
condensing, of course, on any object in the room which
is below the dew point of the air from which the warm
goods came. This may seem like a finely spun theory,
but the writer has had some experience which amply
justifies this explanation. That the moist vapor given
off by the warm goods does not show in the form of
beads of water, or fog, or steam, is no proof that it
does not exist. If the extremes of temperature are
as great as 25° F. condensation will occur on nine
days in ten during the egg storing season. The goods
already in storage are raised in temperature materially by placing in warm goods, which is harmful to
some degree. The logical deduction from above
seems to indicate that warm goods should not be
placed in a room with goods which have been reduced
to the carrying temperature. A separate room
should be provided for this purpose near the receiving room in which the goods coming in warm may be
cooled to very near the temperature of permanent
storage room. This is a refinement which small
houses cannot afford, and which most of the larger
ones do not have.

If you wish to progress compare your results with
those of others. Don't say: "My eggs are as good as

fresh"; test carefully from time to time through the season, and compare quality with those from other houses.

In the foregoing articles I have given my own impressions combined with the data and experience received from others; but I do not care to be held absolutely to any of the statements made, and reserve the right to progress with the rest of you, and do not consider myself bound by any hard and fast rule.

It should be positively understood that a mere theoretical information on this subject is of only limited assistance; and those who undertake new work are advised to put a man in charge who has had experience with the product which it is proposed to handle in storage, as well as acquaintance with the mechanical details of the plant.

www.ingramcontent.com/pod-product-compliance
Lightning Source LLC
Chambersburg PA
CBHW021953190326
41519CB00009B/1239